AF388898

Advanced Support Technologies in Roadway

Advanced Support Technologies in Roadway

Editors

Guoming Liu
Xiangming Hu

MDPI • Basel • Beijing • Wuhan • Barcelona • Belgrade • Manchester • Tokyo • Cluj • Tianjin

Editors
Guoming Liu
Shandong University of
Science and Technology
China

Xiangming Hu
Shandong University of
Science and Technology
China

Editorial Office
MDPI
St. Alban-Anlage 66
4052 Basel, Switzerland

This is a reprint of articles from the Special Issue published online in the open access journal *Applied Sciences* (ISSN 2076-3417) (available at: https://www.mdpi.com/journal/applsci/special_issues/Advanced_Support_Technologies_Roadway).

For citation purposes, cite each article independently as indicated on the article page online and as indicated below:

LastName, A.A.; LastName, B.B.; LastName, C.C. Article Title. *Journal Name* **Year**, *Volume Number*, Page Range.

ISBN 978-3-0365-5775-5 (Hbk)
ISBN 978-3-0365-5776-2 (PDF)

Contents

About the Editors

Guoming Liu

Master supervisor at Shandong University of Science and Technology. His main research area is shotcrete technology. He has presided over a number of projects, such as the youth fund of the National Natural Science Foundation of China and the youth fund of the Natural Science Foundation of Shandong province. He obtained honors for his excellent doctoral dissertation and for being one of the "excellent graduates" of Shandong province. He won the first and second prizes of the China Coal Industry Science and Technology, the first prize of Safety Science and Technology Progress of the China Work Safety Association, the first prize of the National Coal Industry Teaching Achievement Award, and other awards. He has published more than 20 SCI papers and served as a guest editor of SCI journals, such as *Buildings*, *Applied Sciences*, *Sustainability* and *Advances in Civil Engineering*.

Xiangming Hu

Xiangming Hu is a young expert of the Taishan Scholars, Deputy Director of the State Key Laboratory of Mine Disaster Prevention and Control (Incubation Base), recipient of the Sun Yueqi Youth Science and Technology Award, recipient of the Shandong Provincial Excellent Youth Fund, and recipient of the Outstanding Young Scientific and Technological Talent Award in the field of Mining. He has published more than 50 papers in the *Journal of Hazardous Materials*, *Journal of Cleaner Production*, *Cement and Concrete Composites*, *Fuel*, *Construction and Building Materials*, etc. He has presided over three events of the National Natural Science Foundation of China, one sub-project of the National Key Research and Development Program, one of the Excellent Youth Funds of Shandong Province, one of the Excellent Youth Innovation Team Support Programs of Shandong Province, and one of the Science and Technology Development Programs of Shandong Province.

Preface to "Advanced Support Technologies in Roadway"

In the continuous development and application of underground spaces (such as roads, tunnels and underground caverns), safe and efficient support technology is an important means to maintain the stability and integrity of roadways. With the improvement in occupational health requirements and complex and changeable geological conditions, the application of advanced support technology is particularly important for underground work. Exchange and discussion of advanced tunnel support technology is conducive to the safe and efficient development of underground spaces.

The theme of the book is "Advanced Roadway Support Technology", and it aims to collect and display breakthroughs and high-level research progress of advanced support technology, including advanced cement-based materials for tunnel support, advanced surrounding rock control technology, road dust control methods, advanced tunnel support equipment, and advanced road disaster control theory.

This book contains nine papers (eight research papers and one review paper) covering many fields of advanced support technology, including support materials, disaster prevention of surrounding rock damage, deterioration of support structure, pipeline pumping, and road traffic. Xie et al. reported the influence of different additives on the fresh performance of wet shotcrete. Lapian et al. determined the optimum proportion for the asphalt mixture of modified asbuton with PET plastic waste by response surface methodology. Zhang et al. identified damage evolution law and frequency domain distribution characteristics of the acoustic emission signal of deep granite under triaxial loading. Liu et al. summarized the evolution law of mechanical properties and morphological characteristics of shotcrete microstructures at high temperatures. Useche et al. evaluated the effect of a driver's gender on their intention to use an automatic car through robust testing and bias-correction multi-group structural equation modeling. Ma et al. established a prediction model of the compressive strength of artificial sand concrete by response surface methodology. Ma et al. investigated the two-phase flow characteristics of shotcrete in the pipeline based on the CFD-DEM coupling model and on-site measurement. Alonso et al. assessed the development of intelligent transportation systems and other technologies regarding the promotion of population mobility by investigating 1414 Dominicans aged between 18 and 40. Zhu et al. revealed the internal load transfer behavior and failure mechanism of the bird's nest anchor cable anchored structure.

While advanced support technology provides a guarantee for the development of underground spaces, it still faces many challenges. This book is for scholars whose area of research is roadway support, and other scholars are also welcome to read and discuss.

Guoming Liu and Xiangming Hu
Editors

applied sciences

MDPI

Editorial

Advanced Support Technologies in Roadway

Jiahao Sun [1], Guoming Liu [1,2,*] and Xiangming Hu [1,2]

[1] College of Safety and Environmental Engineering, Shandong University of Science and Technology, Qingdao 266590, China
[2] State Key Laboratory of Mining Disaster Prevention and Control Co-Founded by Shandong Province and Ministry of Science and Technology, Shandong University of Science and Technology, Qingdao 266590, China
* Correspondence: skd995978@sdust.edu.cn

Citation: Sun, J.; Liu, G.; Hu, X. Advanced Support Technologies in Roadway. *Appl. Sci.* **2022**, *12*, 10233. https://doi.org/10.3390/app122010233

Received: 14 September 2022
Accepted: 8 October 2022
Published: 11 October 2022

In the continuous development and application of underground spaces (such as roads, tunnels and underground caverns), safe and efficient support technology is an important means to maintain the stability and integrity of roadways. Advanced support technology includes the development of support materials with superior performance, the improvement and innovation of existing support theory, and the development of mechanical equipment with comprehensive functions. For example, the mechanical strength of shotcrete can be enhanced by adding fibers [1,2], improving work efficiency and reducing safety risks through intelligent mechanical equipment [3,4], analyzing the movement characteristics of materials and the control of pipe jam during long distance pumping [5,6], and studying the support structure under complex geological conditions such as high in situ stress [7,8]. Therefore, with the improvement of occupational health requirements and complex and changeable geological conditions, the application of advanced support technology is particularly important for underground work. Exchange and discussion of advanced tunnel support technology is conducive to the safe and efficient development of underground spaces.

This Special Issue aims at collecting and displaying breakthroughs and high-level research progress of advanced support technology, including advanced cement-based materials for tunnel support, advanced surrounding rock control technology, road dust control method, advanced tunnel support equipment, and advanced road disaster control theory.

This Special Issue has published nine papers (eight research papers and one review paper) covering many fields of advanced support technology, including support materials, disaster prevention of surrounding rock damage, deterioration of support structure, pipeline pumping, and road traffic. Xie et al. [9] reported the influence of different additives on the fresh performance of wet-shotcrete, finding that free grout effect and ball effect would significantly affect the performance of wet-shotcrete. The authors discussed the relationship between rheological characteristics and pumpability and spraying ability, and finally put forward suggestions for a mixing ratio meeting the requirements of various shotcrete. Lapian et al. [10] determined the optimum proportion for the asphalt mixture of the modified asbuton with PET plastic waste by response surface methodology, and the final properties of the mixture (stability, Marshall quotient, void in MIX, void mineral aggregate and density) met the specifications of the local engineering department. Zhang et al. [11] identified damage evolution law and frequency domain distribution characteristics of acoustic emission signal of deep granite under triaxial loading. The structure defined signals with amplitude greater than 85 dB, peak frequency greater than 350 kHz and frequency center of gravity greater than 275 kHz as early warning signals of rock failure. Liu et al. [12] summarized the evolution law of mechanical properties and morphological characteristics of microstructures of shotcrete at high temperature, introducing a multi-dimensional morphological formula in the process of heat conduction. The authors established a heat transfer model with spiral shape, and finally put forward opinions regarding the challenges faced by high temperature shotcrete. From the perspective of human factors, Useche et al. [13] evaluated the effect of a driver's gender on their intention to use an automatic car through robust testing and bias-correction multi-group structural equation modeling. They found that the driver's intention to use an automatic car could be explained differently according to their gender, which provided a guiding result for the people those studying traffic planning and road safety. Ma et al. [14] established a prediction model of the compressive strength of artificial sand concrete by response surface methodology, and studied the influence of stone powder, fly ash, and silica powder on the compressive strength of artificial sand concrete. The research found that the content of stone powder has the greatest influence on the compressive strength, while the content of silicon powder has the least effect on the compressive strength. Ma et al. [15] investigated the two-phase flow characteristics of shotcrete in the pipeline based on the CFD-DEM coupling model and on-site measurement, determining the velocity of shotcrete materials and the pressure distribution in the pipeline. The authors found that turbulence and secondary flow prevented the pipeline from blocking, and revealed that the energy loss of aggregate particles in the elbow of the pipeline was about 30 times that in the horizontal straight pipe. Alonso et al. [16] assessed the development of intelligent transportation systems and other technologies regarding the promotion of population mobility by investigating 1414 Dominicans aged between 18 and 40, and proposed the need to strengthen the information and communication flow of emerging transportation-related technologies. Zhu et al. [17] revealed the internal load transfer behavior and failure mechanism of the bird's nest anchor cable anchored structure through theoretical analysis, bird's nest anchor cable pull-out test, and particle flow program numerical simulation test. The authors proposed that the failure mode was the combination of the interface debonding and sliding and rock shear failure.

The submission period of this Special Issue has ended. While advanced support technology provides a guarantee for the development of underground spaces, it still faces many challenges, such as the deterioration of support structure under extreme environment, the impact of complex in situ stress, and the implementation of carbon peak policy. Regarding advanced support technology, the future research directions may face the following aspects. The first aspect is the intellectualization and integration of support technology to realize the integration of spraying technology and the intellectualization of spraying equipment, such as the research and application of shotcrete manipulator and shotcrete robot, which can greatly improve work efficiency and ensure the personal safety of construction workers. The second aspect is the development and application of long-distance pumping in underground spaces. When long-distance pumping is carried out, it is extremely important to ensure the fluidity of materials in the pumping pipeline and prevent pipe plugging. For example, the addition of an air entraining agent can significantly improve the fluidity and workability of concrete, but the migration law and time-varying effect of bubbles in fresh concrete still need to be solved. The third aspect is the development of low-carbon concrete. The implementation of carbon neutralization and carbon peak policies has made cement and related industries face tremendous reforms, such as developing and promoting low-carbon cement materials and improving carbon capture, utilization, and storage technologies.

Author Contributions: Writing—original draft preparation, J.S.; investigation, writing—review and editing, G.L.; supervision, methodology, X.H. All authors have read and agreed to the published version of the manuscript.

Funding: This research received no external funding.

Acknowledgments: Thanks to all the authors and peer reviewers for their valuable contributions to this Special Issue 'Advanced Support Technologies in Roadway'. I would also like to express my gratitude to all the staff and people involved in this Special Issue.

Conflicts of Interest: The authors declare that they have no known competing financial interest or personal relationships that could have appeared to influence the work reported in this paper.

References

1. Kwon, K.Y.; Yoo, D.Y.; Han, S.C.; Yoon, Y.S. Strengthening effects of sprayed fiber reinforced polymers on concrete. *Polym. Compos.* **2015**, *4*, 722–730. [CrossRef]
2. Yoo, D.Y.; Yoon, Y.S.; Banthia, N. Flexural response of steel-fiber-reinforced concrete beams: Effects of strength, fiber content, and strain-rate. *Cem. Concr. Compos.* **2015**, *11*, 84–92. [CrossRef]
3. Liu, G.; Sun, X.; Liu, Y.; Liu, T.; Li, C.; Zhang, X. Automatic spraying motion planning of a shotcrete manipulator. *Intell. Serv. Robot.* **2021**, *1*, 115–128. [CrossRef]
4. Lin, X.; Song, D.; Qin, M.; Zhang, W.; He, X.; Xie, B. An Automatic Tunnel Shotcrete Robot. In Proceedings of the 2019 Chinese Automation Congress (CAC), Hangzhou, China, 22–24 November 2019.
5. Hazaree, C.; Mahadevan, V. Single stage concrete pumping through 2.432 km (1.51 miles): Weather and execution challenges. *Case Stud. Constr. Mater.* **2015**, *12*, 56–69. [CrossRef]
6. Ke, G.; Wang, J.; Tian, B. Simulation Analysis of Pumping and Its Variability for Manufactured Sand Concrete. *ACI Mater. J.* **2019**, *6*, 35–42. [CrossRef]
7. Zhang, H.J.; Li, H.Y.; Wang, X.; Li, H.W. Surrounding rock deformation mechanism of high ground stress soft rock roadways under joint mining influence and corresponding support approach. *Electron. J. Geotech. Eng.* **2015**, *1*, 12109–12119.
8. Peng, R.; Meng, X.; Zhao, G.; Ouyang, Z.; Li, Y. Multi-echelon support method to limit asymmetry instability in different lithology roadways under high ground stress. *Tunn. Undergr. Space Technol.* **2020**, *10*, 3681. [CrossRef]
9. Xie, J.; Cui, X.; Guo, N.; Liu, G. Influence of Mix Proportions on Rheological Properties, Air Content of Wet Shotcrete—A Case Study. *Appl. Sci.* **2021**, *11*, 3550. [CrossRef]
10. Lapian, F.E.; Ramli, M.I.; Pasra, M.; Arsyad, A. The Performance Modeling of Modified Asbuton and Polyethylene Terephthalate (PET) Mixture Using Response Surface Methodology (RSM). *Appl. Sci.* **2021**, *11*, 6144. [CrossRef]
11. Zhang, L.; Ji, H.; Liu, L.; Zhao, J. Time–Frequency Domain Characteristics of Acoustic Emission Signals and Critical Fracture Precursor Signals in the Deep Granite Deformation Process. *Appl. Sci.* **2021**, *11*, 8236. [CrossRef]
12. Liu, G.; Zhao, J.; Zhang, Z.; Wang, C.; Xu, Q. Mechanical Properties and Microstructure of Shotcrete under High Temperature. *Appl. Sci.* **2021**, *11*, 9043. [CrossRef]
13. Useche, S.A.; Peñaranda-Ortega, M.; Gonzalez-Marin, A.; Llamazares, F.J. Assessing the Effect of Drivers' Gender on Their Intention to Use Fully Automated Vehicles. *Appl. Sci.* **2021**, *12*, 103. [CrossRef]
14. Ma, H.; Sun, Z.; Ma, G. Research on Compressive Strength of Manufactured Sand Concrete Based on Response Surface Methodology (RSM). *Appl. Sci.* **2021**, *11*, 3506. [CrossRef]
15. Ma, H.; Sun, Z.; Ma, G. Simulation of Two-Phase Flow of Shotcrete in a Bent Pipe Based on a CFD– DEM Coupling Model. *Appl. Sci.* **2021**, *12*, 3506.
16. Alonso, F.; Faus, M.; Tormo, M.T.; Useche, S.A. Useche. Could Technology and Intelligent Transport Systems Help Improve Mobility in an Emerging Country? Challenges, Opportunities, Gaps and Other Evidence from the Caribbean. *Appl. Sci.* **2022**, *12*, 4759. [CrossRef]
17. Zhu, C.; Zhao, W.; Liu, X. Load Transfer Behavior and Failure Mechanism of Bird's Nest Anchor Cable Anchoring Structure. *Appl. Sci.* **2022**, *12*, 6992. [CrossRef]

Article

Assessing the Effect of Drivers' Gender on Their Intention to Use Fully Automated Vehicles

Sergio A. Useche [1,2,*], **María Peñaranda-Ortega** [3], **Adela Gonzalez-Marin** [4] **and Francisco J. Llamazares** [5]

[1] Research Institute on Traffic and Road Safety (INTRAS), University of Valencia, 46022 Valencia, Spain
[2] Spanish Foundation for Road Safety (FESVIAL), 28004 Madrid, Spain
[3] Department of Basic Psychology and Methodology, University of Murcia, 30100 Murcia, Spain; mariap@um.es
[4] Economic and Legal Sciences, University Center of Defense, 30720 Murcia, Spain; adelaglez@cop.es
[5] Department of Technology, ESIC University, Pozuelo de Alarcón, 28223 Madrid, Spain;
    javier.llamazares@esic.university
* Correspondence: sergio.useche@uv.es

Abstract: Although fully automated vehicles (SAE level 5) are expected to acquire a major relevance for transportation dynamics by the next few years, the number of studies addressing their perceived benefits from the perspective of human factors remains substantially limited. This study aimed, firstly, to assess the relationships among drivers' demographic factors, their assessment of five key features of automated vehicles (i.e., increased connectivity, reduced driving demands, fuel and trip-related efficiency, and safety improvements), and their intention to use them, and secondly, to test the predictive role of the feature' valuations over usage intention, focusing on gender as a key differentiating factor. For this cross-sectional research, the data gathered from a sample of 856 licensed drivers (49.4% females, 50.6% males; $M = 40.05$ years), responding to an electronic survey, was analyzed. Demographic, driving-related data, and attitudinal factors were comparatively analyzed through robust tests and a bias-corrected Multi-Group Structural Equation Modeling (MGSEM) approach. Findings from this work suggest that drivers' assessment of these AV features keep a significant set of multivariate relationships to their usage intention in the future. Additionally, and even though there are some few structural similarities, drivers' intention to use an AV can be differentially explained according to their gender. So far, this research constitutes a first approximation to the intention of using AVs from a MGSEM gender-based approach, being these results of potential interest for researchers and practitioners from different fields, including automotive design, transport planning and road safety.

Keywords: vehicle automation; features; fully automated cars; Multi-Group Structural Equation Modeling (MGSEM); gender; intention; drivers; roadway technologies

Citation: Useche, S.A.; Peñaranda-Ortega, M.; Gonzalez-Marin, A.; Llamazares, F.J. Assessing the Effect of Drivers' Gender on Their Intention to Use Fully Automated Vehicles. *Appl. Sci.* **2022**, *12*, 103. https://doi.org/10.3390/app12010103

Academic Editors: Guoming Liu and Xiangming Hu

Received: 10 November 2021
Accepted: 17 December 2021
Published: 23 December 2021

Publisher's Note: MDPI stays neutral with regard to jurisdictional claims in published maps and institutional affiliations.

1. Introduction

1.1. Automated Vehicles: What Could Drive People to "Make the Shift"?

Nowadays, it is widely known that vehicle-related technologies constitute a core focus to increase safety, efficiency and sustainability of mobility. Accordingly, several technological improvements aimed at supporting a safer and easier driving experience have been developed during the last few decades (e.g., ADAS and other active/passive safety improvements), bringing the automotive market closer and closer to full automation [1,2], thus progressively increasing the SAE level of the vehicles available on the market, looking ahead to the next decade in which, for the case of European countries, about 30% of them are expected to be fully automated (SAE level 5) vehicles [3].

Further, most of the prospective sources on the matter agree on the fact that (just like in any other market) users' perceptions and attitudes play a crucial role for the future of automated vehicles (AVs) and their related transportation dynamics, even though the available empirical information in this regard remains considerably limited [4,5].

However, and as a consequence of different technological, safety and mass communication-related constraints, some studies have argued that the intention to shift to fully automated vehicles could decrease if potential users do not attribute enough value to their different features, especially those related to safety, efficiency and stability [6]. An example of it is the considerable number of non-specialized sources often under or overstressing the actual capabilities of automated vehicles that, far from improving their market-transforming possibilities, can negatively influence both potential consumers' perceptions, same as the willingness of policymakers and transport planners to further invest on adapting road infrastructures as a way to enhance a safer, more sustainable and cost-effective mobility [1,4].

Among all the potential improvements that automated vehicles may represent for their potential users (yet drivers of "conventional" vehicles), there are some features that might be critical for influencing the intention to get involved in this new trend of transport dynamics. For instance, literature emphasizes key aspects such as information flows, ease of driving, energy efficiency, travel swiftness, and perceived safety as essential user-related perceptions to consider with the aim of fostering a more holistic and participatory transition towards automated driving [1,5,6]. Concretely, five of them were addressed in this study.

1.2. Greater Connectivity: Networking Mobility

Among all the social and mobility needs that are expected to be, at least partly, fulfilled through technological developments, recent studies have highlighted mobility networking as one of the core features to be offered by automated vehicles [7]. Instead of operating in a standalone mode, automated vehicles (whose functional basis largely lies on connectivity) are expected to compose cooperative networks useful for different tasks such as traffic control, flow and density monitoring, alerts on critical events and dynamic (real-time) accident prevention [8]. In other words, and apart from maximizing the capacity of urban roads to a substantial extent [9], automated cars are expected to improve "connected mobility" through different resources, including collaborative driving systems (CDS) and vehicle-to-vehicle communication (V2V) systems, thus improving the information available both for our car and for the others' vehicles to make accurately safe decisions [10,11].

Notwithstanding, and although the forecasts about safety, efficiency and technical improvements that having a "more connected" mobility through data sharing would entail, some studies have questioned the growing concerns of drivers in relation to key issues such as data privacy, network stability and reliability, and the possibility that their vehicles (or the networks to which they are connected) could be "hacked" by third parties and, consequently, their privacy and security would get threatened [1,12–14]. Precisely, a recent systematic review found that the behavioral intention to shift to an automated car can be substantially affected by technological fears which are becoming relatively common in current times [15].

1.3. Reduction of Driving Demands

Traditionally, one of the biggest concerns for road safety and vehicle design-related stakeholders (including researchers, designers and practitioners) has been the excessive amount of both physical and psychological demands that the task of driving implies [2,16], especially in long-haul contexts, such as the case of professional driving [17,18]. This critical issue has been linked to different negative outcomes such as fatigue, stress and physical strain, that at the same time remain as reliable predictors of traffic crashes [19–21]. In fact, recent empirical studies have argued about the problematic role of recurrent driving demands also outside the field of professional driving. For instance, factors as common as time pressure may lead drivers to make unsafe decisions and to perform risky driving behaviors, increasing their likelihood to suffer crashes [16,22,23].

Precisely, one of the key benefits of vehicle automation is the progressive reduction of driving demands. For instance, both observational and self-report-based studies have already documented how SAE 2 level-incorporated ADAS (Advanced Driving Assistance

Systems) reduce the number and degree of many (often simultaneous) demands to which drivers are subjected during their everyday trips [14,24].

However, it is worth mentioning that results in these aforementioned regards remain considerably inconsistent, whereas other researchers have argued that not all the potential users of automated vehicles would, actually, have a very positive valuation of the fact that a machine might take decisions and execute driving tasks on behalf of them, especially in cases such as: (i) they have a high preference for driving themselves, (ii) they consider that their driving skills can be better than anyone else's, often also over any machine, (iii) they either enjoy the experience of assuming sensations related to risk and speed, (iv) they can be afraid of losing their jobs or experiencing unexpected changes as a consequence of vehicle automation, and/or (v) they simply prefer "staying in control" of their cars, especially at safety-critical moments and complex driving scenarios [25–27].

1.4. Fuel/Energy Consumption Saving

One of the most commonly featured benefits of vehicle automation is vehicle energy efficiency [28], explaining an improved potential to transform energy-related dynamics in transport, thus increasing their contribution to both environmental sustainability (including greenhouse emissions) and economy [29,30]. In other words, it is expected a widespread adoption of automated cars might reasonably reduce air pollution and benefit many stakeholders, including drivers, passengers and other road users, especially when involved in large-use transport spheres and services, such as taxis and private hire cars (PHCs) [31].

Notwithstanding, recent studies highlight that, even though the projections are really promising in environmental terms, there are still many uncertainties prevailing around the actual operation of fully (SAE level 5) automated vehicles, as this basically remains a hypothesized technology [31,32], whose specific features should substantially vary during the next few years, also in consideration of different factors and dynamics that can be transformed (e.g., the price of different forms of energy, the ruling ones, the new technologies that could be discovered or massified and the storage capacity of batteries and their efficiency, in the case of electric cars) in the term of the next few years [33–35].

At the (potential) user level, however, advances in energy terms and their subsequent monetary savings are usually a relevant feature for decision-making, as has been evidenced in previous studies related to vehicle automation [28,36,37]. Indeed, sustainability-related settings are nowadays considered as a critical part of both the consolidation of automated (and clean) driving as a transportation pattern, and the consumer preferences of today's drivers, given the high level of social discussion on the subject and awareness of the issue that exists in most industrialized countries [37–39].

1.5. Travel Efficiency

Another key feature that should be addressed in relation to vehicle automation is their overall hypothesized substantial contribution to reduce costs and travel times [25,36]. Especially under conditions of high demand, automated vehicles and connected transport technologies are expected to help substantially reduce the average number of minutes a driver or passenger spends on each of their trips [40,41]. In a recent study, Sonnletiner, Friedrick and Richter determined that factors such as the improvement of current algorithms through artificial intelligence developments and machine learning, and the increasing number of units (connected cars) in urban traffic networks, despite experiencing some difficulties in their early implementation phases, could significantly optimize the trips and the time used for them in a few years [42].

Nevertheless, and regardless of objective estimations, the latest stated preference studies in these regards have shown how people's perceptions (that are theorized to exert an effect on demand rates) remain relatively skeptical in regard to the actual extent to which, in a near future, highly and fully automated vehicles (i.e., SAE levels 4 and 5) could, indeed, improve urban dynamics related to travel time and trip efficiency [43].

1.6. Improved Safety

Finally, the role of safety in vehicle automation must be addressed. Since a couple decades ago, different theoretical and empirical sources have argued that human knowledge and ability, even provided with greater autonomy, decision-making capacity and discernment, they are considerably lower when compared with most of the technologies commonly used in the field of vehicle automation [9,11]. Therefore, it could be expected that, in a relatively controlled environment (e.g., the so-called smart cities), the diversification of automated cars would help to reduce the number of errors, traffic violations and their subsequent accidents, injuries and deaths, in addition to all the subsequent costs that the aforementioned issues commonly involve [41].

At the user level, it is nowadays virtually impossible to provide rigorously suitable estimations on to what extent road safety will become benefited from the implementation and widespreading of automated vehicles, especially in absence of estimations about the extent to which drivers' behavior will remain as a core crash predictor [6,15]. However, some initial studies have determined that safety-related perceptions of nowadays' drivers might influence their intention to shift towards higher SAE level vehicles, but also that demographic differences of potential users (especially as for gender) might contribute to identify differences in the development of attitudes, perceptions and intentions towards automated cars and their driving assistance features [1,6,43].

1.7. Study Aims and Hypotheses

Bearing in mind the aforementioned considerations and insights provided by previous literature, the two aims of this study were: first, to assess the relationships among drivers' demographic factors, their valuation of five key features of automated vehicles and the intention to use a fully automated AV in the future. In this regard, it was hypothesized that drivers' assessment of these features—that fully automated cars are expected to have—might be significantly related to their self-reported intention to switch to AVs.

Secondly, this study also aimed at testing the predictive role of the assessment given to these five features on drivers' intention to use automated cars, focusing on gender as a potentially differentiating factor. As for this second study aim, it was hypothesized that AV feature-based assessments would have a significant (but differential) effect over usage intention, i.e., there will exist structural differences in the explanation of the intention to use an automated vehicle depending on whether there is a male or a female driver.

2. Methods

2.1. Design and Study Setting

With the aim of providing a methodological overview to the readers, the steps of this cross-sectional study, successively described in the different subsections of the methods, are graphically synthesized in Figure 1.

Given the current COVID-19-related social distancing protocols, this cross-sectional research was performed through an electronic survey written in Spanish, distributed to an approximate number of 1550 individuals included in a pre-existing mailing list shared among universities and research centers during the first half of the year 2021. Potential participants included in the mailing list (whose basic features are overall similar to the general Spanish population, at least in terms of gender and age), were recruited through a convenience sampling method, receiving a personal invitation to partake in this study, stating its objective, participation dynamics and ethical considerations surrounding it.

The only two inclusion criteria were to be a currently licensed non-professional driver (regardless of the type of vehicle) residing in Spain, and to read and accept the conditions included in the Informed Consent form before starting to respond the questionnaire, which was mandatory. Additionally, it is worth saying that participants were briefly contextualized in SAE levels prior to responding the questionnaire, with the aim to differentiate the concepts of "partially-to-high automated vehicle" (SAE levels 2 to 4), and "fully automated vehicle" (SAE level 5).

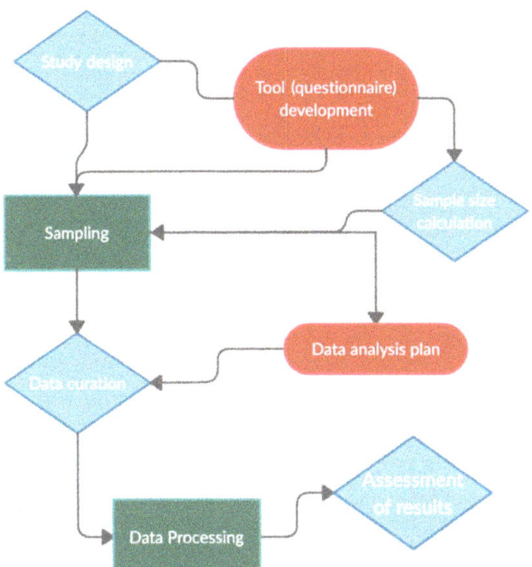

Figure 1. Data Flow Chart (DFC) describing the methodological steps of the study.

Regarding other sample-related issues, the response rate was about 56% of the invited partakers; no economical rewards or stimulus were offered to participants. In order to achieve an acceptable degree of representativeness: (i) we assumed the full Spanish census of drivers (about 27 million drivers for the year 2020) [44] as population size, and (ii) although population representativeness can be only partially assumed on the basis of a non-probabilistic sampling method, an a priori calculation of the minimum sample size was carried out using the following formula:

$$\left(S = z^2 \sigma \ / \ e^2 \right) \tag{1}$$

where S represents the sample size; z the standardized value for different levels of confidence, e.g., $z = 1.96_{(95\%)}$, or $2.58_{(99\%)}$, σ represents the standard deviation (commonly set as 0.50), and e represents the maximum error allowed, with $\alpha = 0ñ0.05_{(5\%)}$. The resulting number suggested a minimum of about 690 subjects (better if proportionally distributed by sex and age), assuming a maximum margin of error of 5% ($\alpha = 0.05$) and a beta (β) of 0.20, which allows for an 80% power. It is worth saying that, although this implies the need of retrieving more cases, it lessens the margin of error of the study.

The average time required to partake in the research (i.e., responding to the electronic survey) was about 8 min. In order to avoid potentially biased responses, before starting the survey, it was emphasized that the data would be exclusively used for statistical research purposes and their participation was anonymous.

2.2. Study Sample

This study analyzed the data obtained from a sample of $n = 856$ licensed drivers aged between 18 and 65, with a mean of $M = 40.05$ ($SD = 11.47$) years. Participants were residents of all the 17 Autonomous Communities (Regions) of Spain, with a sample proportion between 4–9% per region. From the study sample, 49.4% of the drivers were females and 50.6% were males. Table 1 presents detailed demographic characteristics of the study participants.

Table 1. Study partakers' sociodemographic data and basic driving features.

Feature	Category	Frequency	Percentage	Population Census Data [a]
Gender	Female	423	49.4%	50.9%
	Male	433	50.6%	49.1%
Age Group	<25	92	10.7%	8.2% [b]
	25–34	197	23%	13.0%
	35–44	253	29.5%	17.4%
	45–54	223	26%	18.7%
	>54	91	10.6%	42.7%
Educational level	Primary school	63	7.4%	
	Secondary-high school	376	43.9%	
	Technical education	133	15.5%	
	University	284	33.2%	
Years since firstly licensed	Less than 5 years	102	11.9%	
	5 to 10 years	138	16.1%	
	More than 10 years	616	72%	
Type of vehicle (most driven)	Private car	801	93.5%	
	Motorcycle	32	3.7%	
	Van/Light truck	14	1.6%	
	Other (moped, electric two-seater)	9	1.2%	
Driving frequency (weekly basis)	Daily	376	44%	
	4–6 days a week	239	27.9%	
	2–3 days a week	141	16.6%	
	1 day a week or less	100	11.7%	

Notes for the Table: [a] Population-based data [44] was included to compare sample distribution in terms of gender and age group; [b] age group percentages are based on the population census over 18, to make it comparable with driving population.

2.3. Description of the Questionnaire

The research questionnaire was composed of four main sections:

The first section inquired about demographic data, including gender (male/female/other; "other" was never chosen as a response, so that the variable was dichotomized), educational level, city of residence, income level and driving-related information (years since firstly licensed, type of vehicle most commonly driven, driving frequency).

The second part of the survey comprised a Likert-based questionnaire consisting of 20 questions, using a 5-point (1–5) scale, i.e., from 1 (not at all → no relevance/value/improvement perceived at all) to 5 (too much → much relevance/value/improvement perceived). Drivers were asked about their assessment of five different key characteristics of fully automated vehicles, namely: greater connectivity (4 items; $\alpha = 0.726$), reduced driving demands (4 items; $\alpha = 0.705$), energy efficiency (4 items; $\alpha = 0.769$), travel efficiency (4 items; $\alpha = 0.762$), and improved safety (4 items; $\alpha = 0.801$). The valuation of these features was based on a four-item subscale for each feature, addressing: (a) perceived relevance; (b) value attributed; (c) the degree to which it may improve driving experience, and (d) the actual viability or likelihood of the feature.

The third part of the survey comprised a 5-item ($\alpha = 0.840$) attitude questionnaire aimed at assessing drivers' intention to use an automated vehicle (example item: If during the next years I will have enough budget, I plan to buy an AV). The scale used a (1–5) scale, being 1 (totally disagree) and 5 (totally agree) [6].

Finally, the fourth part of the survey was aimed to assess drivers' degree of interaction with Information and Communication Technologies (ICTs; to be used as a control variable) in a scale ranging between 1 (very scarce interaction) and 5 (very high interaction). The item used for this purpose was: "Please tell us your degree of usual interaction with smartphones, computers and/or other devices which are normally connected to the Internet".

2.4. Data Processing

After performing basis statistical procedures, i.e., descriptive statistics, bivariate Spearman's r_s or *rho* correlations and Brown–Forsythe's mean comparisons (robust tests, as basic normality and homoscedasticity-related assumptions were not met), a gender-based Multi-Group Structural Equation Model (MGSEM) was built up. For this purpose, there were used Bootstrap-based robust maximum likelihood estimations (i.e., 10,000 bootstrap samples and 95% confidence intervals), in order to handle non-normality issues, as most of the study variables did not meet the basic assumption of univariate normality, and multivariate normality was not met either, as usually happens in self-report-based studies [45]. The model fit was evaluated by using Chi-square (χ^2), Normed Fit Index (NFI), Incremental Fit Index (IFI), Comparative Fit Index (CFI), and Root Mean Square Error of Approximation (RMSEA) [18].

As for the punctual features of the model used to test the hypothesized structural relationships among measured variables, the multivariate relationships between female and male drivers' demographic/psychosocial factors and their intention to use an AV, it was composed of the six exogenous variables and one endogenous factor that are shown in Section 3.3. This is statistically more accurate than separately testing genders as separate populations since it considers the full sample parameters for fitting the models. The direct effects of the model, their confidence intervals (at the level 95%) and significance levels were calculated following the bootstrap method, specifically through a Monte Carlo (parametric) procedure, favoring that, e.g., the results of the estimates may be bias-corrected, do not present problems of normality, type I errors (false positives) in regression paths can be avoided, and constitutes a reasonable alternative to other estimation methods such as Satorra–Bentler or Weighted Least Square Mean and Variance adjusted (WLSMV). For this study, SEM modeling tasks were performed with SPSS AMOS software (Version 26.0; IBM Corp., Armonk, NY, USA).

Estimators were calculated controlling for income level, degree of interaction with ICTs and driving experience. According to the specialized literature [18,46], it is commonly accepted, as rules of thumb, that a set of CFI/NFI/IFI coefficients greater than 0.900 and a Root Mean Square Error of Approximation lower than 0.080 (better if <0.060; [18]), plus the coherence of the model data with its theoretical assumptions, constitute insights of an acceptable model fit to the data. When possible, the model's fit was improved taking into account the largest and more theoretically parsimonious modification indexes.

3. Results

3.1. Descriptive Data

Table 2 shows the mean values and non/parametric bivariate correlations of the variables measured in the study, divided in three blocks: participants' demographic data, their assessment of five automated vehicle features, and their intention to use them.

Spearman's (*rho*) correlation coefficients can be interpreted similarly to Pearson's (*r*) coefficients, ranging between −1 (very strong negative association) and 1 very strong positive association, where 0 implies there is no relation between two variables. Overall, demographic variables, and especially age, have shown interesting associations with the valuation of two of these features. Concretely, age has been found significantly (and negatively) associated with the valuation of AVs' greater connectivity and reduced driving demands. In other words, the higher the drivers' age, the lesser is their valuation of these two AV features. Further, it draws attention to how drivers with a greater degree of interaction to Information and Communication Technologies (ICTs) tend to value to a greater extent travel efficiency and increased safety features, but also to self-report a greater intention to shift to an automated vehicle, also raising the need of including age and interaction with ITCs among the control variables to perform inferential models on the intention to use AVs (see Section 3.3).

Table 2. Basic descriptive data and bivariate (Spearman's *rho*) correlations between study variables.

	Variable	Mean [a]	SD [b]	1	2	3	4	5	6	7	8
				Demographic data							
1	Age	40.05	11.47	– [c]							
2	Educational Level	– [c]	–	−0.117 **	–						
3	Interaction with ICTs	2.82	0.98	−0.214 **	0.191 **	–					
				Assessment of five automated car features							
4	Greater connectivity	3.25	1.98	−0.109 **	0.060	0.003	–				
5	Reduced driving demands	1.90	1.67	−0.124 **	0.025	0.051	−0.025	–			
6	Fuel/energy consumption saving	3.42	1.01	0.036	0.044	0.067	−0.161 **	−0.105 **	–		
7	Travel efficiency	3.29	0.94	0.021	0.026	0.114 **	−0.207 **	−0.148 **	0.457 **	–	
8	Increased safety	2.63	0.62	0.043	0.071 *	0.075 *	−0.285 **	−0.203 **	0.391 **	0.506 **	
				Intention to use an automated car							
9	Intention	2.80	0.60	0.058	0.099 **	0.030	−0.184 **	0.004	0.306 **	0.394 **	0.481 **

Notes for the Table: [a] Average value for the full sample; [b] SD= standard deviation; [c] value cannot be computed (applies to purely ordinal variables or the correlation of a variable with itself, where 1.0 is an invariant value); * correlation is significant at the $p < 0.050$ level (2-tailed); ** correlation is significant at the $p < 0.010$ level (2-tailed).

As for the bivariate relationships between the intention to use AVs and the five features covered by this study, it was found that higher scores on the assessment of a greater connectivity and reduced driving demands are negatively correlated with the valuation of fuel saving, travel efficiency and safety features. On the other hand, the bivariate correlations between increased safety, trip efficiency and lower energy consumption assessments remain positive among them, same as with the intention to use AVs. In other words, it seems that the improvements developed in these three terms might be (among the five) those that are, indeed, potentially associated with a greater intention to shift to an automated vehicle in the near future.

3.2. Gender-Based Differences

After assessing bivariate correlations between these three groups of variables, descriptive gender differences were explored. Given that the assumption of normality was not met in the case of most variables used in the study, especially because Likert questionnaires have an ordinal nature, and variances were rather heteroscedastic, robust (Brown– Forsythe's F) tests were used for this purpose. Unlike traditional ANOVA tests, this technique uses a different denominator for the "F" equation, adjusting the mean square through the observed variances of each group, instead of dividing by the mean square of the error. The results of mean comparison tests are fully available in Table 3, being readable and interpretable in the same way as One-way Analysis of Variance tests.

Overall, significant differences could be established between three out of the five AV features addressed in the study, namely: Fuel/energy consumption saving, Travel efficiency and Increased safety, where valuations were always higher among male drivers. On the other hand, females: (i) self-report a greater level of interaction with ICTs than males and (ii) tend to value to a greater extent the AVs' greater connectivity and driving demands reduction, but these differences do not reach the cut-off points needed for assuming significance at a 95% level of confidence.

3.3. Structural Analyses

Based on the aforementioned theoretical assumptions of the study, the effect of gender over the extent to which male and female drivers intend to use an AV was examined through a MGSEM (Multi-Group Structural Equation Modeling) approach, that differs from using gender as a dummy category within a structural model encompassing other predictive variables, whose effects can be hypothesized to differ in nature according to drivers' gender. Instead, it allows differentially assessing the effect of the exogenous factors on the dependent variable for each group, making it possible to compare the "mechanisms" by which these relations can be explained for the case of each gender.

Table 3. Descriptive statistics and mean comparisons by driver's gender.

Variable	Group	Descriptives		Brown–Forsythe Test			
		Mean	SD	Statistic [a]	df1 [b]	df2 [c]	Sig.
Interaction with ITCs	Males	2.75	0.94	4.921	1	847.620	<0.001 ***
	Females	2.90	1.01				
Greater connectivity	Males	3.11	2.00	3.259	1	653.656	0.072 [N/S]
	Females	3.39	1.96				
Reduced driving demands	Males	1.81	1.61	1.705	1	653.961	0.192 [N/S]
	Females	1.98	1.73				
Fuel/energy consumption saving	Males	3.52	1.02	9.785	1	853.793	0.008 **
	Females	3.30	0.98				
Travel efficiency	Males	3.44	0.95	23.882	1	853.707	<0.001 ***
	Females	3.13	0.91				
Increased safety	Males	2.73	0.70	21.911	1	850.655	<0.001 ***
	Females	2.52	0.64				
Intention to use	Males	2.86	0.60	6.413	1	852.986	0.035 *
	Females	2.76	0.61				

Notes for the Table: [a] Asymptotically F distributed; [b] df1= B–F test degrees of freedom 1; [c] df2 = B–F test degrees of freedom 2; * significant at the $p < 0.050$ level (2-tailed); ** significant at the $p < 0.010$ level (2-tailed); *** significant at the $p < 0.001$ level (2-tailed); [N/S] non-significant difference.

In this sense, data were split into two gender-based groups (i.e., reference categories): a group of 423 (49.4%) female, and a group of 433 (50.6%) male drivers, both of them with acceptable sample size and proportionality for a comparative examination. Utilizing multi-group (MGSEM) analysis, the hypothesized structural model was adjusted to control for demographic and driving-related differences, and to fit the data according to gender, at the same time considering the parameters of the full sample.

The resulting Structural Equation Model, simultaneously fitted for both gender groups ($x^2_{(13)}$ = 73.220, $p < 0.001$; NFI = 0.917; CFI = 0.926; IFI = 0.930; RMSEA = 0.077, CI 90%= 0.061–0.095), is presented through two merged graphical models in Figure 2. Qualitatively, the magnitude and significance levels of paths from exogenous variables to self-reported rates show differential trends between male and female drivers.

Table 4. Multi-group SEM model to predict drivers' intention to use automated vehicles.

Group A: Male Drivers						
	Path		S.P.C. [a]	S.E. [b]	C.R. [c]	p [d]
Intention	←	Interaction with ICTs	0.086	0.04	2.102	0.036 *
Intention	←	Greater connectivity	−0.106	0.096	−2.191	0.028 *
Intention	←	Reduction of driving demands	0.058	0.118	1.201	0.230 [N/S]
Intention	←	Fuel/energy consumption saving	0.104	0.045	2.229	0.026 *
Intention	←	Travel efficiency	0.231	0.013	4.751	<0.001 ***
Intention	←	Increased safety	0.301	0.05	6.128	<0.001 ***
Group B: Female Drivers						
	Path		S.P.C.	S.E.	C.R.	p
Intention	←	Interaction with ICTs	0.017	0.045	0.417	0.677 [N/S]
Intention	←	Greater connectivity	0.048	0.099	1.02	0.308 [N/S]
Intention	←	Reduction of driving demands	0.157	0.112	3.315	<0.001 ***
Intention	←	Fuel/energy consumption saving	0.056	0.049	1.173	0.241 [N/S]
Intention	←	Travel efficiency	0.166	0.014	3.241	0.002 **
Intention	←	Increased safety	0.397	0.056	7.941	<0.001 ***

Notes for the Table: [a] S.P.C.= Standardized path coefficient; [b] S.E.= standard error; [c] C.R. = critical ratio; [d] bias-corrected p-value; * = path is significant at the level $p < 0.050$; ** = path is significant at the level $p < 0.010$; *** = path is significant at the level $p < 0.001$; [N/S] non-significant path.

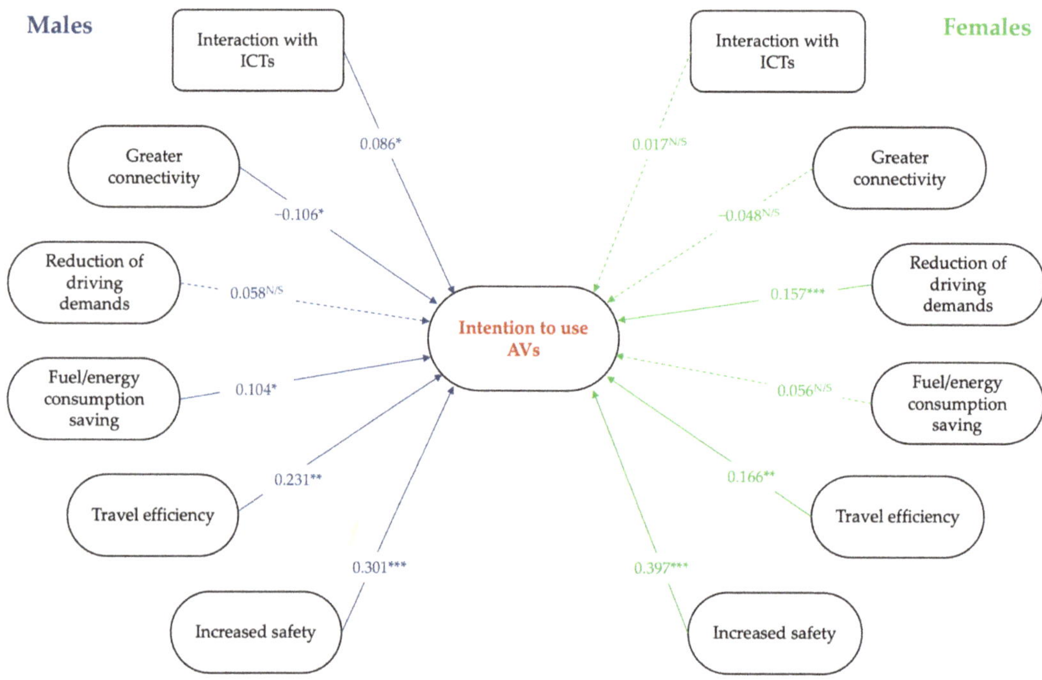

Figure 2. Multi-group structural model showing standardized path coefficients for self-reported intention to use AVs: male drivers (**left**) and female drivers (**right**). $^{N/S}$ Non-significant path (discontinuous arrows); * $p < 0.050$; ** $p < 0.010$; *** $p < 0.001$. Additional information on other (miscellaneous) statistical parameters is available in Table 4.

The Standardized Path Coefficients (SPCs) and their significance values (see solid lines in Figure 2 and the right column of Table 4) in the bias-corrected MGSEM suggest that there exist both structural similarities and differences, as follows:

In regard to structural similarities, the MGSEM results show only two akin path relationships between male and female drivers, namely: (i) drivers' valuation of trip efficiency, and (ii) safety improvements of automated vehicles significantly predict their intention to use them.

On the other hand, and as for structural differences, it was found that: (i) drivers' degree of interaction with Information and Communication Technologies (ITCs) are a significant predictor of intention among males, but it remains non-significant for the case of female drivers; (ii) unlike the case of females, a greater connectivity seems to significantly decrease the self-reported intention among male drivers; (iii) reduction of driving demands is a significant predictor only among female drivers, and (iv) fuel/energy consumption saving significantly increases intention to use AVs only among male drivers.

4. Discussion

The core aims of this study were, first, to assess the relationships among drivers' demographic factors, the valuation given to five key features of automated vehicles, and their intention to use AVs, and secondly, to test the predictive role of the valuation given to these five features on drivers' intention, focusing on gender as a key differentiating factor. Therefore, the findings in each one of these regards will be discussed below, in consideration of both the study hypotheses and the existing literature on the matter.

In regard to the first aim, it was initially hypothesized that drivers' valuations of five different features greatly known about fully-automated vehicles might be significantly related to their self-reported intention to switch to AVs. Based on the data provided by this relatively extensive sample of 856 drivers, this study found that, indeed, there are bivariate relationships between drivers' intentions to shift to an automated vehicle and their valuation of five of its most relevant features, within those presented prospectively by the literature [3,5,9]. In this regard, it is worth emphasizing that, in a period of approximately 10 years (for when it is expected to be a large enough market in European countries), some other relevant features and/or technological developments could substantially affect how automated vehicles are perceived, and how transport dynamics could be transformed as a consequence of their potential impacts in different spheres [7,40,47].

Further, the descriptive results of this study suggest some interesting points worthwhile to be briefly discussed: firstly, that AV feature-related valuations may largely differ on the basis of gender (to be discussed below, when addressing the second study aim), and secondly, the existence of statistically significant differences in the self-rated interaction with ICTs by gender, in which case women present significantly higher scores than their male counterparts. As for this finding, some recent studies have found that, regardless of gender gaps in terms of accessibility in many countries, but especially in emerging economies [48], females might tend to keep a greater mean daily interaction with some "connected devices" in terms of frequency, intensity, and even as for subsequent problems derived from it [49,50], even though findings remain somewhat inconsistent across different literature sources.

The second aim of this study (i.e., to test the predictive role of these five key features in a gender-based perspective) allowed to a priori hypothesize that that AV-feature valuations would have a significant (but still differential) effect over drivers' intentions to use them. Overall, the results found endorsed the hypothesis that there exist certain multi-group structural [51] differences in the explanation of the intention to use an automated vehicle based on its differential features, at least between male and female drivers.

Although this multivariate technique had not been ever applied to the study of users' preferences, attitudes and/or intentions over automated vehicles (AVs), previous researches have already analyzed the role of gender as a key potential determinant and (in other cases) mediator between, demographic/psychosocial issues and perceptual and behavioral outcomes of several groups of road users (for a summary, please see: [52–54]).

Furthermore, gender differences constitute a "traditional" but underused issue in road safety and transport planning studies. With a certain frequency, gender-focused studies are solely descriptive and ignore the covariant and "invisible" relationships and effects among multiple factors, as happens in simple mean comparisons that (although still interesting), do not properly delve into the mechanisms by which multiple gender differences can simultaneously interact [51,53]. Therefore, and beyond the significant mean differences found in these terms, this study provides an in-depth analysis of the differential role of various features expected to be found in SAE 5 level cars between drivers of both genders, serving as a starting point for further research on the matter.

Addressing each one of the six variables included in the significant MGSEM model as predictors of drivers' intention to use AVs, a summary is presented below:

- *Interaction with ICTs*. This technology demand-related variable significantly explained males' (although not females') intention in a positive way. In other words, and although the magnitude of the significant path coefficient is low, it can be assumed that male drivers keeping a greater degree of contact with technological devices might also tend to have a greater intention to shift from a conventional car to an AV. Previously, some few studies have assessed the link between interaction with ICTs and the intention of using cars provided with higher SAE level technologies, also finding a predictive (and positive) link amongst them [6,55]. Additionally, and although no gender focus had never been applied before to this specific issue, it is worth remarking how other studies have linked previous interactions between people and technology with their acceptance of new technological solutions and improvements [56,57].

- *Greater connectivity of AVs:* Same as Interaction with Information and Communication Technologies, this feature of AVs had a structurally different role, in function to driver's gender. Namely, it negatively influenced intentions of male drivers, although not of females. In this regard, previous research endorses the role of information-related privacy and security as a prevalent concern for many of both technology and road users, including the case of drivers [1,9,12,14]. This is, in other words, an issue that (besides all current ethical discussions existing in these regards [13]) might need additional efforts aimed at (i) guaranteeing users a desirable level of security, stability and reliability [58], and (ii) communicating what are the real aspects on which the connectivity of the AVs will be used to their powerful users, in a clear, participatory and comprehensive manner [13,55,56].

- *Reduction of driving demands:* Contrary to what is observed in the case of the *greater connectivity* feature, the fact that automated cars reduce driving demands seems to represent "an incentive" (to a significant extent) only for women, among whom the predictive relationship between the feature and their intention to use AVs remains positive. In this regard, a reasonably plausible explanation is the fact that, overall, males tend to believe they perform significantly better than others in different tasks, including vehicle driving [59,60], as well as being more optimistic and perceiving themselves as less prone to get in a crash, particularly when judging their own driving skills [60]. Moreover, and as vehicle automation aims at progressively decreasing the number and complexity of the demands imposed to the driver by trip-related tasks [9], and fully automated cars are designed to practically drive themselves, it is not surprising to find emerging concerns surrounding this (nowadays) set of almost unknown dynamics in transportation [27,42], especially if it is taken into account that the most complex levels of automation still remain practically unexploited on public roads in the eyes of drivers [7,16,28].

- *Fuel/energy consumption saving:* This feature, apart from depicting another structural difference in regard to AV use-related intention between male and female drivers, constituted a significant predictor only for the case of males, and the association was positive. In other words, it suggests that the extent to which an automated vehicle might be energy efficient might represent a potential issue affecting decision-making of male drivers, even besides what concerns environmental sustainability settings [29,32]. Consistently, few studies have argued that, apart from sustainability-related factors [30,59,61], fuel efficiency is a feature often pursued by drivers on the basis of an economical motivation. However, what draws more attention is the fact that it is men who, in general, show that they are more informed about fuel-efficient driving, when compared with females, as concluded by McIlroy and Stanton in a recent applied study on eco-driving knowledge performed in the United Kingdom [62].

- *Travel efficiency and increased safety:* In these two cases, the relationship is positive and similar in terms of magnitude. Therefore, no gender differences were found in the case of age, which showed to have a similar influence in both spheres. In other words, and coherently with previous researches performed on these two key features, this study's outcomes suggest that efficiency and safety constitute crucial issues for vehicle automation, at least in the eye of potential customers [6,14,58]. Therefore, the findings consistently suggest that the greater a driver's valuation of AV's trip efficiency and safety, the higher is the extent to which they are predisposed to move closer to vehicle automation for their trips [1,55], regardless of gender.

5. Conclusions

5.1. Study Outcomes and Implications

Findings from this work suggest that: firstly, drivers' demographic factors, their assessment of five features of automated vehicles (i.e., increased connectivity, reduced driving demands, fuel and trip-related efficiency, and safety improvements) keep a significant set of multivariate relationships with their intention to use them in the future.

Secondly, and even though there are few structural similarities, the intention of drivers to use an AV can be differentially explained according to their gender, being males' intentions are affected by connectivity, fuel consumption, energy and trip efficiency, and safety-related issues, whereas females' intention to use an AV remains, rather, linked to driving demands, trip efficiency, and safety features. In brief, these differential attitudinal trends could differentiate both their preferences and potential practices related to autonomous driving in the near future.

Finally, and as for the practical implications of this study, this research constitutes the first approximation made so far to the intention of using AVs from a gender-based MGSEM approach, being these results of potential interest for many stakeholders, including:

- Firstly, researchers and practitioners on road safety can find relevant insights about (among others) the relationship between perceived safety and drivers' intention to use vehicles whose features might help decreasing their crash likelihood.
- Secondly, automotive design-related stakeholders may find it useful to find (with good empirical proof) that there are, indeed, several key differences in relation to the factors enhancing the intention to use AVs between male and female drivers, which can be used for product development purposes.
- Finally, one of the key issues highlighted in this study is the strategic value of automated cars for both sustainability and transport efficiency. Therefore, transport planners may find in this study a quick overview of some gender-based gaps in the valuation of e.g., energy and trip efficiency issues of AVs, which would be worth addressing in the near future in both planning, communication and training processes.

5.2. Limitations of the Study and Further Research

Although this research analyzed the data from a relatively extensive sample of drivers, which was also considerably balanced in terms of age and gender, and the essential statistical and theoretical assumptions (included Goodness-of-Fit criteria) were met, it is worth acknowledging some essential limitations and technical shortcomings that could have biased the research outcomes, so that readers will make a careful interpretation of the data presented by this study.

First of all, an anonymous interview does not fully deter common method biases (CMBs) in responses, especially if there are addressed topics related to the social discussion, such as vehicle automation and transportation dynamics. Secondly, and although based on a relatively extensive literature review, the set of AV-related features measured by this study is only partial: only a few of them, even though all which were relevant were covered

Secondly, there are many other factors potentially affecting drivers' attitudes and intentions towards AVs that remain unexplored in this research.

Thirdly, the authors consider it might be suggestible to complementarily acquire further insights on this interesting issue by means of e.g., in-depth interviews, repeated measures over time (as this is a very changing process) and mixed research methods, with the aim to maximize the explanation of gender-based differences in regard to user-related issues affecting future transportation dynamics and trends.

Author Contributions: Conceptualization, S.A.U. and M.P.-O.; methodology, S.A.U.; software, F.J.L. and M.P.-O.; validation, S.A.U., A.G.-M. and F.J.L.; formal analysis, S.A.U.; investigation, S.A.U., M.P.-O. and A.G.-M.; resources, F.J.L.; data curation, A.G.-M. and M.P.-O.; writing—original draft preparation, S.A.U. and A.G.-M.; writing—review and editing, S.A.U., M.P.-O. and A.G.-M.; visualization, F.J.L. and M.P.-O.; supervision, S.A.U. and F.J.L.; project administration, S.A.U. All authors have read and agreed to the published version of the manuscript.

Funding: This study was supported by the Spanish Foundation for Road Safety (FESVIAL) with grant RG/002/2021 (Grant recipient: S.A.U.).

Institutional Review Board Statement: The study was conducted according to the guidelines of the Declaration of Helsinki, and approved by the Ethics Committee of the Research Institute on Traffic and Road Safety at the University of Valencia (protocol code HE0002150621 from 15 June 2021).

Informed Consent Statement: Informed consent was obtained from all subjects involved in the study.

Data Availability Statement: The data presented in this study are available upon reasonable request from the corresponding author. The data are not publicly available given that the development of other research products derived from this project is still underway.

Acknowledgments: The authors would like to thank all our partakers, research staff members, and institutional stakeholders for their active involvement in this study. A special mention to Boris Cendales for the reading proof of the final version of the manuscript.

Conflicts of Interest: The authors declare no conflict of interest regarding this manuscript.

References

1. Lijarcio, I.; Useche, S.A.; Llamazares, J.; Montoro, L. Availability, Demand, Perceived Constraints and Disuse of ADAS Technologies in Spain: Findings from a National Study. *IEEE Access* **2019**, *7*, 129862–129873. [CrossRef]
2. Roh, C.G.; Kim, J.; Im, I.J. Analysis of Impact of Rain Conditions on ADAS. *Sensors* **2020**, *20*, 6720. [CrossRef] [PubMed]
3. McKinsey & Co. Automotive Revolution—Perspective towards 2030: How the Convergence of Disruptive Technology-Driven Trends Could Transform the Auto Industry. 2016. Available online: http://go.uv.es/G6b8o63 (accessed on 10 November 2021).
4. Hryniewicz, K.; Grzegorczyk, T. How different autonomous vehicle presentation influences its acceptance: Is a communal car better than agentic one? *PLoS ONE* **2020**, *15*, e0238714. [CrossRef] [PubMed]
5. Huang, T. Research on the use intention of potential designers of unmanned cars based on technology acceptance model. *PLoS ONE* **2021**, *16*, e0256570. [CrossRef]
6. Montoro, L.; Useche, S.A.; Alonso, F.; Lijarcio, I.; Bosó-Seguí, P.; Martí-Belda, A. Perceived safety and attributed value as predictors of the intention to use autonomous vehicles: A national study with Spanish drivers. *Saf. Sci.* **2019**, *120*, 865–876. [CrossRef]
7. Malik, S.; Khan, M.A.; El-Sayed, H. Collaborative Autonomous Driving-A Survey of Solution Approaches and Future Challenges. *Sensors* **2021**, *21*, 3783. [CrossRef] [PubMed]
8. Tan, H.; Zhao, F.; Hao, H.; Liu, Z. Evidence for the Crash Avoidance Effectiveness of Intelligent and Connected Vehicle Technologies. *Int. J. Environ. Res. Public Health* **2021**, *18*, 9228. [CrossRef] [PubMed]
9. Hancock, P.A.; Nourbakhsh, I.; Stewart, J. On the future of transportation in an era of automated and autonomous vehicles. *Proc. Natl. Acad. Sci. USA* **2019**, *116*, 7684–7691. [CrossRef]
10. Ghazal, T.M.; Said, R.A.; Taleb, N. Internet of vehicles and autonomous systems with AI for medical things. *Soft. Comput.* **2021**, 1–13. [CrossRef]
11. Hancock, P.A. Driving into the Future. *Front. Psychol.* **2020**, *11*, 574097. [CrossRef]
12. Linkov, V.; Zámečník, P.; Havlíčková, D.; Pai, C.W. Human Factors in the Cybersecurity of Autonomous Vehicles: Trends in Current Research. *Front. Psychol.* **2019**, *10*, 995. [CrossRef]
13. Othman, K. Public acceptance and perception of autonomous vehicles: A comprehensive review. *AI Ethics* **2021**, *1*, 355–387. [CrossRef]
14. Lijarcio, I.; Useche, S.A.; Llamazares, J.; Montoro, L. Perceived benefits and constraints in vehicle automation: Data to assess the relationship between driver's features and their attitudes towards autonomous vehicles. *Data Brief* **2019**, *27*, 104662. [CrossRef]
15. Keszey, T. Behavioural intention to use autonomous vehicles: Systematic review and empirical extension. *Transp. Res. Part C Emerg. Technol.* **2020**, *119*, 102732. [CrossRef]
16. Liang, D.; Lau, N.; Baker, S.A.; Antin, J.F. Examining Senior Drivers' Attitudes Toward Advanced Driver Assistance Systems After Naturalistic Exposure. *Innov. Aging* **2020**, *4*, igaa017. [CrossRef]
17. Daza, I.G.; Bergasa, L.M.; Bronte, S.; Yebes, J.J.; Almazán, J.; Arroyo, R. Fusion of optimized indicators from Advanced Driver Assistance Systems (ADAS) for driver drowsiness detection. *Sensors* **2014**, *14*, 1106–1131. [CrossRef] [PubMed]
18. Useche, S.A.; Alonso, F.; Cendales, B.; Llamazares, J. More Than Just "Stressful"? Testing the Mediating Role of Fatigue on the Relationship Between Job Stress and Occupational Crashes of Long-Haul Truck Drivers. *Psychol. Res. Behav. Manag.* **2021**, *14*, 1211–1221. [CrossRef] [PubMed]
19. Serrano-Fernández, M.J.; Boada-Grau, J.; Robert-Sentís, L.; Vigil-Colet, A.; Assens-Serra, J. Predictive power of selected factors over driver stress at work. *Int. J. Occup. Saf. Ergon.* **2021**, *27*, 416–424. [CrossRef]
20. Lemke, M.K.; Apostolopoulos, Y.; Hege, A.; Sönmez, S.; Wideman, L. Understanding the role of sleep quality and sleep duration in commercial driving safety. *Accid. Anal. Prev.* **2016**, *97*, 79–86. [CrossRef]
21. Useche, S.A.; Ortiz, V.G.; Cendales, B.E. Stress-related psychosocial factors at work, fatigue, and risky driving behavior in bus rapid transport (BRT) drivers. *Accid. Anal. Prev.* **2017**, *104*, 106–114. [CrossRef] [PubMed]
22. Čulík, K.; Kalašová, A. Statistical Evaluation of BIS-11 and DAQ Tools in the Field of Traffic Psychology. *Mathematics* **2021**, *9*, 433. [CrossRef]
23. Čulík, K.; Kalašová, A.; Kubíková, S. Simulation as an Instrument for Research of Driver-vehicle Interaction. *MATEC Web. Conf.* **2017**, *134*, 8. [CrossRef]
24. De-Las-Heras, G.; Sánchez-Soriano, J.; Puertas, E. Advanced Driver Assistance Systems (ADAS) Based on Machine Learning Techniques for the Detection and Transcription of Variable Message Signs on Roads. *Sensors* **2021**, *21*, 5866. [CrossRef]

25. Teoh, E.R.; Kidd, D.G. Rage against the machine? Google's self-driving cars versus human drivers. *J. Saf. Res.* **2017**, *63*, 57–60. [CrossRef] [PubMed]
26. Pettigrew, S.; Fritschi, L.; Norman, R. The Potential Implications of Autonomous Vehicles in and around the Workplace. *Int. J. Environ. Res. Public Health* **2018**, *15*, 1876. [CrossRef]
27. Nees, M.A. Acceptance of Self-driving Cars: An Examination of Idealized versus Realistic Portrayals with a Self- driving Car Acceptance Scale. *Proc. Hum. Factors Ergon. Soc. Annu. Meet.* **2016**, *60*, 1449–1453. [CrossRef]
28. Yuen, K.F.; Chua, G.; Wang, X.; Ma, F.; Li, K.X. Understanding Public Acceptance of Autonomous Vehicles Using the Theory of Planned Behaviour. *Int. J. Environ. Res. Public Health* **2020**, *17*, 4419. [CrossRef]
29. Brown, K.E.; Dodder, R. Energy and emissions implications of automated vehicles in the U.S. energy system. *Transp. Res. D Transp. Environ.* **2019**, *77*, 132–147. [CrossRef]
30. Massar, M.; Reza, I.; Rahman, S.M.; Abdullah, S.M.H.; Jamal, A.; Al-Ismail, F.S. Impacts of Autonomous Vehicles on Greenhouse Gas Emissions-Positive or Negative? *Int. J. Environ. Res. Public Health* **2021**, *18*, 5567. [CrossRef]
31. Freedman, I.G.; Kim, E.; Muennig, P.A. Autonomous vehicles are cost-effective when used as taxis. *Injury Epidemiol.* **2018**, *5*, 24. [CrossRef]
32. Ghadikolaei, M.A.; Wong, P.K.; Cheung, C.S.; Zhao, J.; Ning, Z.; Yung, K.F.; Wong, H.C.; Gali, N.K. Why is the world not yet ready to use alternative fuel vehicles? *Heliyon* **2021**, *7*, e07527. [CrossRef] [PubMed]
33. Backhaus, R. Battery Raw Materials—Where from and Where to? *ATZelectron. Electron. Worldw.* **2021**, *16*, 38–43. [CrossRef]
34. Weiss, M.; Zerfass, A.; Helmers, E. Fully electric and plug-in hybrid cars—An analysis of learning rates, user costs, and costs for mitigating CO_2 and air pollutant emissions. *J. Clean. Prod.* **2019**, *212*, 1478–1489. [CrossRef] [PubMed]
35. Kampker, A.; Offermanns, C.; Heimes, H.; Bi, P. Meta-analysis on the Market Development of Electrified Vehicles. *ATZ Worldw.* **2021**, *123*, 58–63. [CrossRef]
36. Lajunen, T.; Sullman, M.J.M. Attitudes Toward Four Levels of Self-Driving Technology Among Elderly Drivers. *Front. Psychol.* **2021**, *12*, 682973. [CrossRef] [PubMed]
37. Mora, L.; Wu, X.; Panori, A. Mind the gap: Developments in autonomous driving research and the sustainability challenge. *J. Clean. Prod.* **2020**, *275*, 124087. [CrossRef]
38. Rousseau, S.; Deschacht, N. Public Awareness of Nature and the Environment During the COVID-19 Crisis. *Environ. Resour. Econ.* **2020**, *76*, 1149–1159. [CrossRef] [PubMed]
39. Burns, L.D.; Jordan, W.C.; Scarborough, B.A. *Transforming Personal Mobility*; The Earth Institute, University of Columbia: New York, NY, USA, 2013.
40. Auld, J.; Sokolov, V.; Stephens, T.S. Analysis of the effects of connected-automated vehicle technologies on travel demand. *Transp. Res. Rec.* **2017**, *2625*, 1–8. [CrossRef]
41. Childress, S.; Nichols, B.; Charlton, B.; Coe, S. Using an activity-based model to explore the potential impacts of automated vehicles. *Transp. Res. Rec.* **2015**, *2493*, 99–106. [CrossRef]
42. Sonnleitner, J.; Friedrich, M.; Richter, E. Impacts of highly automated vehicles on travel demand: Macroscopic modeling methods and some results. *Transportation* **2021**. [CrossRef]
43. De Almeida Correia, G.H.; Looff, E.; van Cranenburgh, S.; Snelder, M.; van Arem, B. On the impact of vehicle automation on the value of travel time while performing work and leisure activities in a car: Theoretical insights and results from a stated preference survey. *Transp. Res. Part A Policy Pract.* **2019**, *119*, 359–382. [CrossRef]
44. INE. *Population Figures/Basic Demographic Indicators. Web Resource*; Instituto Nacional de Estadística: Madrid, Spain, 2020. Available online: https://www.ine.es/dyngs/INEbase/es/operacion.htm?c=Estadistica_C&cid=1254736176951&menu=ultiDatos&idp= 1254735572981 (accessed on 15 December 2021).
45. Byrne, B. *Structural Equation Modeling with AMOS. Basic Concepts, Applications and Programming*, 2nd ed.; Routledge Taylor & Francis Group: New York, NY, USA, 2010.
46. Marsh, H.W.; Hau, K.T.; Wen, Z. In search of golden rules: Comment on hypothesis-testing approaches to setting cutoff values for fit indexes and dangers in overgeneralizing Hu and Bentler's (1999) findings. *Struct. Equ. Model. Multidiscip. J.* **2004**, *11*, 320–341. [CrossRef]
47. García, A.; Camacho-Torregrosa, F.J. Influence of Lane Width on Semi- Autonomous Vehicle Performance. *Transp. Res. Rec. J. Transp. Res. Board* **2020**, *2674*, 279–286. [CrossRef]
48. Garcia, D.; Mitike Kassa, Y.; Cuevas, A.; Cebrian, M.; Moro, E.; Rahwan, I.; Cuevas, R. Analyzing gender inequality through large-scale Facebook advertising data. *Proc. Natl. Acad. Sci. USA* **2018**, *115*, 6958–6963. [CrossRef]
49. Baumann, E.; Czerwinski, F.; Reifegerste, D. Gender-Specific Determinants and Patterns of Online Health Information Seeking: Results from a Representative German Health Survey. *J. Med. Internet Res.* **2017**, *19*, e92. [CrossRef]
50. Chen, B.; Liu, F.; Ding, S.; Ying, X.; Wang, L.; Wen, Y. Gender differences in factors associated with smartphone addiction: A cross-sectional study among medical college students. *BMC Psychiatry* **2017**, *17*, 341. [CrossRef]
51. Rezapour, M.; Ksaibati, K. Application of multi-group structural equation modelling for investigation of traffic barrier crash severity. *Int. J. Injury Control Saf. Promot.* **2020**, *27*, 232–242. [CrossRef]
52. Chen, H.Y.; Donmez, B. What drives technology-based distractions? A structural equation model on social-psychological factors of technology-based driver distraction engagement. *Accid. Anal. Prev.* **2016**, *91*, 166–174. [CrossRef]

53. Useche, S.A.; Hezaveh, A.M.; Llamazares, F.J.; Cherry, C. Not gendered . . . but different from each other? A structural equation model for explaining risky road behaviors of female and male pedestrians. *Accid. Anal. Prev.* **2021**, *150*, 105942. [CrossRef]
54. Useche, S.A.; Montoro, L.; Alonso, F.; Tortosa, F.M. Does gender really matter? A structural equation model to explain risky and positive cycling behaviors. *Accid. Anal. Prev.* **2018**, *118*, 86–95. [CrossRef]
55. Nordhoff, S.; Stapel, J.; van Arem, B.; Happee, R. Passenger opinions of the perceived safety and interaction with automated shuttles: A test ride study with 'hidden' safety steward. *Transp. Res. Part A Policy Pract.* **2020**, *138*, 508–524. [CrossRef]
56. Habibovic, A.; Lundgren, V.M.; Andersson, J.; Klingegård, M.; Lagström, T.; Sirkka, A.; Fagerlönn, J.; Edgren, C.; Fredriksson, R.; Krupenia, S.; et al. Communicating Intent of Automated Vehicles to Pedestrians. *Front. Psychol.* **2018**, *9*, 1336. [CrossRef] [PubMed]
57. Woods, D.D. Commentary designs are hypotheses about how artifacts shape cognition and collaboration. *Ergonomics* **1998**, *41*, 168–173. [CrossRef]
58. Ahangar, M.N.; Ahmed, Q.Z.; Khan, F.A.; Hafeez, M. A Survey of Autonomous Vehicles: Enabling Communication Technologies and Challenges. *Sensors* **2021**, *21*, 706. [CrossRef]
59. Al-Balbissi, A.H. Role of Gender in Road Accidents. *Traffic Inj. Prev.* **2003**, *4*, 64–73. [CrossRef]
60. Ring, P.; Neyse, L.; David-Barett, T.; Schmidt, U. Gender Differences in Performance Predictions: Evidence from the Cognitive Reflection Test. *Front. Psychol.* **2016**, *7*, 1680. [CrossRef] [PubMed]
61. Muro-Rodríguez, A.I.; Perez-Jiménez, I.R.; Gutiérrez-Broncano, S. Consumer Behavior in the Choice of Mode of Transport: A Case Study in the Toledo-Madrid Corridor. *Front. Psychol.* **2017**, *8*, 1011. [CrossRef]
62. McIlroy, R.C.; Stanton, N.A. What do people know about eco-driving? *Ergonomics* **2017**, *60*, 754–769. [CrossRef]

applied sciences

Article

Could Technology and Intelligent Transport Systems Help Improve Mobility in an Emerging Country? Challenges, Opportunities, Gaps and Other Evidence from the Caribbean

Francisco Alonso [1], Mireia Faus [1], Maria T. Tormo [1] and Sergio A. Useche [2,*]

[1] INTRAS (Research Institute on Traffic and Road Safety), University of Valencia, 46022 Valencia, Spain; francisco.alonso@uv.es (F.A.); mireia.faus@uv.es (M.F.); m.teresa.tormo@uv.es (M.T.T.)

[2] ESIC Business & Marketing School, 46021 Valencia, Spain

* Correspondence: sergioalejandro.useche@esic.edu

Citation: Alonso, F.; Faus, M.; Tormo, M.T.; Useche, S.A. Could Technology and Intelligent Transport Systems Help Improve Mobility in an Emerging Country? Challenges, Opportunities, Gaps and Other Evidence from the Caribbean. *Appl. Sci.* **2022**, *12*, 4759. https://doi.org/10.3390/app12094759

Academic Editors: Guoming Liu and Xiangming Hu

Received: 23 April 2022
Accepted: 7 May 2022
Published: 9 May 2022

Publisher's Note: MDPI stays neutral with regard to jurisdictional claims in published maps and institutional affiliations.

Abstract: Apart from constituting a topic of high relevance for transport planners and policymakers, support technologies for traffic have the potential to bring significant benefits to mobility. In addition, there are groups of "high potential" users, such as young adults, who constitute an essential part of the current market. Notwithstanding, and especially in low and middle-income countries (LMICs), their knowledge and acceptance remain understudied. This study aimed to assess the appraisal of intelligent transport systems (ITS) and other technological developments applicable to mobility among Dominican young adults. Methods: In this study, we used the data gathered from 1414 Dominicans aged between 18 and 40, responding to the National Survey on Mobility in 2018 and 2019. Results: Overall, and although there is a relatively high acceptance, attributed value, and attitudinal predisposition towards both intelligent transportation systems and various support technologies applicable to mobility, the actual usage rates remain considerably low, and this is probably exacerbated by the low and middle-income status of the country. Conclusions: The findings of this study suggest the need to strengthen information and communication flows over emerging mobility-related technologies and develop further awareness of the potential benefits of technological developments for everyday transport dynamics.

Keywords: technology; mobility; intelligent transport systems; perception; low and middle-income countries; Dominican Republic

1. Introduction

Intelligent transportation systems (ITS) can be understood as the set of applications, devices, and technological systems that facilitate traffic management, control, and monitoring [1,2]. Among their multiple advantages, ITSs contribute to improving users' safety [3], traffic flows [4], and trip-related decisions [5]. In practical terms, this means minimizing congestion, reducing pollution, and (equally important) enhancing mobility as a potentially positive and safe experience [6,7].

There is great variability of ITS systems that have been applied on roads around the world for some years [8]. Intelligent mobility enables connection between vehicles (V2V) and between vehicles and infrastructure (V2I) [9]. Thus, the number of transport means equipped with devices capable of detecting risks and facilitating driving maneuvers for drivers keeps rising [10]. In addition, many urban areas have also incorporated electronic devices, such as cameras or smart traffic lights, helping cities enforce traffic laws and protecting vulnerable road users [11,12].

In addition, ITSs are already present in the market through products, such as advanced driving assistance systems (ADAS) and autonomous cars, both useful to minimize human errors during driving and improve overall road safety [13,14]. Therefore, at the user level, a great acceptance of technologies applied to traffic among their potential consumers could be expected. However, recent studies suggest that opinions, intentions, and usage willingness of ITSs could be rather split, especially among users who currently have minute information about them [15].

1.1. New Technologies in Traffic: A Socially Challenging Issue?

In general, emerging technologies in mobility are well-valued by most of the existing studies on traffic safety [16], with young men being the population group expressing the most positive opinions [17]. Specifically, the most highly valued aspects are safety improvements [18], reduced traffic congestion and travel time [19], amenities for the elderly and/or physically challenged [20,21], environment and fuel savings [22,23], and other miscellaneous benefits [24,25].

However, several studies have pointed out that the geographic provenance of their users could be the main factor affecting the assessment of ITS systems [26,27]. Namely, people living in highly industrialized countries with high rates of economic wealth are favorable to the inclusion of traffic-related technology. In contrast, people living in emerging countries, where the evidence is scarce, have been implied to remain more reluctant [26]. Enhanced by many social disparities, one of the factors influencing these gaps might be the current unequal penetration of information and communication technologies (ICTs) among different countries.

Precisely, one of the regions that seems more affected by these issues is the Caribbean, whose countries are in the lower-middle zone of the world ranking, which reflects the number of households with internet access [28]. In contrast, this list is headed mainly by European and Asian countries [29]. Conversely, almost all the existing studies conducted in low and middle-income countries (LMICs) show a lower predisposition to accept and use advanced support technologies in traffic [27,30]. Some of the main constraints commonly expressed by users from LMICs are the fear of potential cybersecurity issues or system failures [31], personal data treatment [32], a feeling of excessive monitoring [33,34], and environmental concerns [35].

From a theoretical point of view, different approaches have been developed to understand interpersonal differences in the degree of acceptance of technology. Initially, Schifter and Ajzen (1985) developed the theory of planned behaviour (TPB) [36], which shows that the intention to behave, modulated by attitude and subjective norm, may be the best behavioral predictor. A few years later, Davis (1989) proposed the technology acceptance model (TAM) [37], in which perceived ease of use and perceived usefulness are hypothesized to directly influence attitudes towards technology and, consequently, the intention to use it. Overall, both TPB and TAM models agree on the imperative need for users to have a positive attitude and value a certain behavior as necessary steps to increase its likelihood. However, to the best of our knowledge, these assumptions have never been tested in this region, specifically regarding support technologies for mobility.

1.2. Study Context: The Case of the Dominican Republic

In recent years, the Dominican Republic has made significant progress in the traffic and road safety sector [38]. Specifically, with the creation of the National Institute of Transit and Land Transportation (INTRANT) in 2017, multiple measures have been developed to improve traffic efficiency and reduce crash rates in the country [39]. Concretely, many advances have been observed in terms of transport policy, infrastructural advancements, road safety communication, and a progressive improvement of the automotive fleet [40,41].

Notwithstanding, crash rates continue to be among the highest in the world [42], partly due to the low adherence to policymaking outcomes and the many other social problems relatable with everyday transport dynamics [43]. In addition, the Dominican population, as in other Latin American countries, such as Brazil, Jamaica, and Colombia, tends to privilege private vehicles over massive transport, explaining an abruptly large increase in the use of motorcycles [43,44], precarious small-sized public transport means [45], and other low-cost everyday solutions [46,47].

In this regard, previous studies suggest that bringing people closer to transport-related technology (applicable to everyday mobility) would be helpful to face many of the common threats affecting the Dominican population, starting with young segments of the population, given their greater positive attitudes, experiences and capabilities to interact with ICTs and ITSs [17,38,48]. Although the current developments in this regard are still considerably scarce, it is worth mentioning that, to date, some first technology-related advances for improving transport dynamics are taking place, especially with the implementation of local mobile applications development by the government to improve the mobility of vehicles, as well as a progressive modernization of accessibility and payment methods in public transport (e.g., biometric devices and contactless payment systems) [45,49].

1.3. Study Aim

Bearing in mind the aforementioned considerations, this study aimed to assess the appraisal of intelligent transport systems and other technological developments applicable to mobility among Dominican young adults (aged between 18–40). In this regard, it was hypothesized that this segment of the population would attribute a high degree of value to ITSs and other supporting technological developments for traffic settings, even though demographic factors might modulate these appraisals.

2. Materials and Methods

2.1. Participants

This nationwide study used the information provided by 1411 young adult Dominicans between 18 and 40 years of age. These participants were divided into two samples. The first (n = 661) was gathered in the 2018 national survey on mobility, and the second (n = 753) corresponded to the national survey of 2019, allowing us to make comparisons on the state of affairs in this regard between the two different years (Table 1). As an important issue to be remarked, a sample for the year 2020 could not be gathered due to the COVID-19 pandemic.

Table 1. Sociodemographic characteristics of the two study sub-samples.

Factor	Value	2018		2019	
		n	%	n	%
Gender	Man	308	46.6%	375	49.8%
	Woman	353	24.1%	378	50.2%
	Total	661	100%	753	100%
Habitat	Urban	502	75.9%	611	81.1%
	Rural	159	24.1%	142	18.9%
	Total	661	100%	753	100%
Job situation	Unemployed	138	21.1%	286	38%
	Full-time job	389	56.7%	328	43.3%
	Part-time job	134	20.3%	148	19.7%
	Total	661	100%	753	100%
Driving status	Licensed	236	35.7%	311	41.3%
	Not licensed	425	64.3%	442	58.7%
	Total	661	100%	753	100%

The data used for this study were gathered using a stratified random sampling method. Therefore, distribution was proportional to the population, according to the ONE (National Statistics Office of the Dominican Republic) census, by age, sex, habitat, and province, as shown in Table 1. Regarding sample size, the minimum number of participants was determined to be n = 385 for each year, assuming a confidence level of 95%, a maximum margin of error of 5% (alpha = 0.05), and a beta of 0.20, in order to achieve statistical representativeness. In addition, the sampling characteristics are similar to those employed by previous empirical road safety research in the country [47–49]. Personal data were managed in accordance with current data protection laws, and respondents participated voluntarily and anonymously.

2.2. Design, Procedure and Instruments

As aforementioned, the data used for this study were directly obtained from the National Mobility Survey of the Dominican Republic in 2018 [50] and 2019 [51]. The questionnaire consisted of several blocks of questions that explored citizens' perceptions of various traffic, mobility, and road safety issues. Among these addressed topics were: private and public transport, mobility on foot and by bicycle, knowledge of institutions and traffic laws, and communication campaigns developed in the Dominican Republic. In addition, there was a specific section for the knowledge and valuation of mobility-related technology, focused on ICTs and ITS systems, whose results are discussed in this paper.

The questionnaire was administered through personal interviews. The data collection process was carried out using a CAPI (computer-assisted interviewing) system on recorded and geo-referenced tablets to reduce interviewing time and avoid errors in the recording of data.

To achieve the proposed objectives, the following variables were considered:

- Sociodemographic variables and driving data: sex, age, working situation, habitat, frequency of driving.
- Knowledge of intelligent transportation systems: questions were asked using a dichotomic scale (Yes/No): "Do you know what ITS systems are?"; "Do you know what Smart Mobility is?"; and "Do you know what Smart Cities are?".
- Perceptions about technology on traffic: A Likert scale from 0 to 10 was used for the following questions: "To what extent do you consider that technology (in general) can be useful for the improvement of mobility?"; "To what extent do you consider mobile apps as useful for the improvement of mobility?"; and "To what extent do you consider ITS systems as useful for the improvement of mobility?"
- Use of mobile apps during urban trips: a generic question was asked about the use of apps with two answer options (yes/no), as well as an open question specifying which apps are used. The degree of belief that apps could encourage the use of public transport and the degree to which they could improve traffic and mobility in the Dominican Republic were also evaluated through a Likert scale from 0 to 10.
- As environmental affairs have been acquiring growing importance in mobility and technology-related improvements, a complementary question was included asking for participants' self-reported degree of environmental concern, using a scale ranging from 0 to 10 (see Figure 1).
- Individuals' willingness to use different means of transport (i.e., private means, public transport, or walking) to perform their everyday trips was assessed through a Likert scale from 0 to 10.
- Finally, the importance given to ten different features of intelligent transport systems (e.g., 'traffic information'; see Figure 2 for full list) was assessed through a scale ranging from 0 to 10.

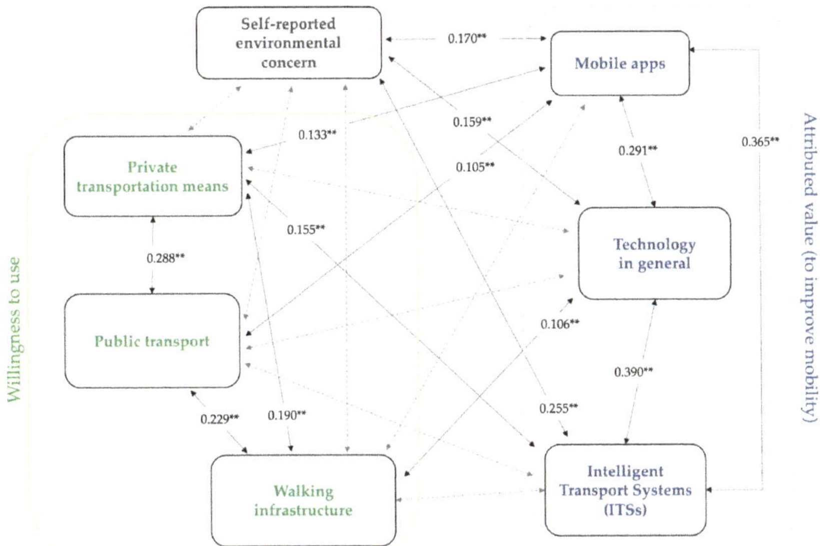

Figure 1. Correlation networks: Pearson's bivariate associations. *Notes:* ** Correlation is significant at the level $p < 0.010$.

Figure 2. Priority of the use of ITS systems according to Dominican citizens.

2.3. Data Processing

This study aimed to describe and characterize the amount of knowledge of ITS systems held by Dominican individuals through descriptive analysis. ANOVA and chi-square tests were also used to determine how the variables changed over time between the two surveys. Pearson correlations and their corresponding graphical network representations have also been performed to establish possible statistical relations among the analyzed variables. All the statistical tests performed used differential significance level criteria of $p < 0.050$, $p < 0.010$, and $p < 0.001$. IBM SPSS (Statistical Package for Social Sciences) version 26.0 (Armonk, NY, USA) was used to perform all statistical analyses.

2.4. Ethics

The Ethics Committee of Research in Social Science in Health of the University of Valencia was consulted before the study. It confirmed that the research met general ethical criteria and agreed with the Declaration of Helsinki (IRB approval number: HE0001251019). After the team gave them a detailed description of the research goal and all preceding considerations, participants consented to participate in the study.

3. Results

Overall, the descriptive results of this study show how the self-reported knowledge of intelligent transport systems remained relatively scarce on the part of young Dominicans in both 2018 and 2019. Despite this, a statistically significant increase in the number of people who claim to know about ITS is observed in this short period of time, being 4.2% in 2018 and 8.2% in 2019 (Table 2).

Table 2. Level of knowledge and perception of ITS systems.

		2018		2019				
		%	n	%	n	Chi^2	p	Phi
ITS systems knowledge	Yes	4.2%	28	8.2%	62	Chi$^2_{(1)}$ = 9.439	0.002	−0.082
	No	95.8%	633	91.8%	691			

	2018			2019				
	X	SD	n	X	SD	n	ANOVA	p
Valuation of technology (in general) as something useful for the improvement of traffic?	7.41	2.42	661	8.03	2.84	753	$F_{(11,412)}$ = 18.953	<0.001

Another interesting finding was the fact that only 3% of participants reported having an awareness of the concept of "Smart Cities" and 4.7% of Smart Mobility. In regard to their valuation of the usefulness of technological improvements for traffic dynamics, young Dominican adults reported similar average values (M = 7.41; SD = 2.42 for 2018 and M = 8.03; SD = 2.84 for 2019), with a slight but still statistically significant increase between these two years, as also shown in Table 2.

3.1. Correlation Analyses

Another issue worth exploring was the set of associations among technological, environmental, and transport-related issues, as perceived by Dominicans. As this is a bivariate procedure, the correlation network presented in Figure 1 graphically depicts the relationships between pairs of them. In brief, some interesting outcomes stand out:

Firstly, and regarding the bivariate relationships between technology-related affairs and environmental concerns, it was found that participants' assessments on (i) technology in general; (ii) apps; and (iii) ITSs for traffic improvement are all positively (and significantly) correlated with their self-reported degree of environmental concern.

Secondly, it was found that technology-related perceptions were associated to a certain extent with participants' willingness to use different means of transportation:

(a) The value attributed to mobile apps for improving mobility was positively correlated to individuals' willingness to use public and private transport means, although not with the intention of using walking infrastructure to perform their daily trips.
(b) On the other hand, the overall valuation of technology for mobility was correlated to usage willingness only in the case of walking trips.
(c) Finally, the value attributed to intelligent transport systems for improving mobility was positively associated with the willingness of individuals to use public transportation, but not with their intention to travel either by private means of transport or walking.

3.2. Which ITS Features Are More Important for Young Dominicans?

As aforementioned, ITS systems were overall valued as positive for improving urban mobility (M = 6.49; SD = 2.76). In this sense, examining the differential level of acceptance of various of their core features seems relevant.

Overall, the most highly rated ITS features were: (i) detection of dangerous locations (M = 6.95; SD = 2.60); and (ii) priority pass for certain vehicles (ambulances, emergency vehicles, public transport, etc.) (M = 6.92; SD = 2.61). On the other hand, the lowermost rated ITS features were parking information (M = 6.47; SD = 2.40) and traffic information (M = 6.49; SD = 2.51).

3.3. Valuation and Usage Level of Mobile Apps in Urban Trips

Participants' valuation of mobile applications both in everyday life and as a tool for invigorating mobility improvement showed a relatively good assessment, as shown in Table 3. Overall, and following the aforementioned [0–10] scale, mobile apps' relevance for everyday life was scored with an M = 7.22/10 in 2018, while this mean value increased to M = 7.49/10 for the year 2019, the difference being significant, and its trend of an increasing nature.

Table 3. Level of perception and use of mobile apps for the improvement of mobility.

Value Attributed to Apps		2018			2019				
		X	SD	n	X	SD	n	ANOVA	p
For enhancing everyday life		7.22	2.59	661	7.49	2.94	753	$F_{(11,412)}$ = 3.370	0.067
For improving mobility		7.34	2.61	661	7.72	3.04	753	$F_{(11,412)}$ = 6.341	0.012
		2018			2019				
		%		n	%		n	Chi^2	p	Phi
Use of mobile apps as	Yes	13.6%		86	23.0%		173	Chi^2(1) = 23.359	<0.001	0.129
a support for urban trips	No	86.4%		575	77%		580			

Secondly, a similar trend was found in regard to the value attributed to mobile apps for enhancing mobility, with a significant increase between 2018 (M = 7.34/10) and 2019 (M = 7.72/10), showing a growing appreciation and perceived usefulness among applications potentially applicable to the traffic environment.

However, and despite this relatively high rating of mobile applications on mobility, their usage level remains scarce for 2019, even though there has been a significant increase of almost 10% in the number of people using them during their urban trips (see Table 3).

Finally, it is worth depicting that the mobile apps relatable to mobility had the highest frequency of use. Figure 3 shows the most used applications, among which those related to navigation, maps, and routes stand out.

Although a very small proportion of respondents (up to 23%) reported having a minimum degree of mobility-related app usage, it is worth remarking that: (i) this trend is, in any case, rising, and (ii) there was a slight increase in the use of apps related to active transportation, from 0.3% in 2018 to 0.5% in 2019.

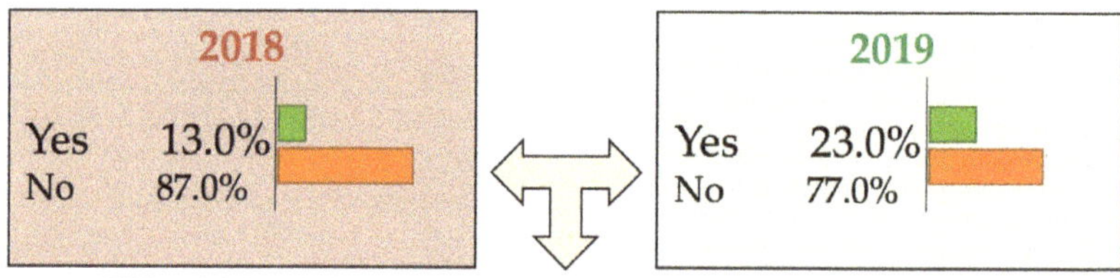

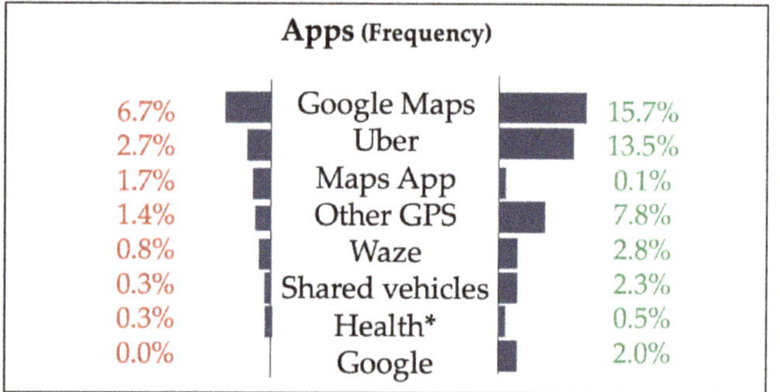

*Only applicable to active trips (walking, cycling, etc.)

Figure 3. Most commonly used apps during urban trips (absolute frequency).

4. Discussion

The core aim of this study was to assess the appraisal of intelligent transport systems and other technological developments applicable to mobility among Dominican young adults (aged between 18–40). Overall, the findings show that there is a relatively high acceptance, attributed value, and attitudinal predisposition. However, the actual usage rates remain considerably low, probably exacerbated by the Low and Middle-Income status of the country.

Despite the relatively low knowledge and usage rates, it is remarkable how awareness of ITS is starting to increase, especially considering several benefits they may imply for both road users and transportation dynamics, ranging from the minimization of traffic jams in urban areas and the reduction of gas and waste pollutants in cities to the prevention of traffic crashes [5,6].

Further, and apart from assessing the self-reported knowledge of a key segment of the Dominican population about ITS systems, it is worth addressing their social perception of the importance of technology for mobility, as well as various potential specific spheres, including those related to mobile apps and connected devices [52].

This growing understanding of technology benefits for mobility could be hypothesized to help enhance potential ITS applications in the country. However, it seems necessary to keep strengthening relevant issues remarked by preceding literature about technology consumption on traffic, such as perceived usefulness, attitudes, safety improvements and personal data management [16,36,37,53]. Thus, the practical value of this research is contributing to the understanding of the current perceptions and beliefs of young Dominicans regarding new technologies and their application for traffic, mobility, and road safety. The conclusions obtained will serve as a starting point for the development of social awareness measures and policies.

4.1. Social Perception of the Role of Technology for Traffic and Mobility Improvement

As described before, self-reported knowledge of ITS systems was rather scarce among Dominicans, as well as other related concepts, such as 'smart mobility' and 'smart cities'.

Previous studies agree on the prevalence of this panorama across several Latin American countries, despite the important role of Information and Communication Technologies (ICTs) in the growth of these territories [54].

For instance, research conducted in Colombia exposes the growing misinformation and number of misbeliefs surrounding the aforementioned concepts [55], even though government lines of action revolve around digitization and technological innovation [56]. Similar conclusions can be drawn from the implementation of some ITS systems in Brazil, where the role of users and institutions in the specific technological adaptation to the problems of each region is emphasized [57,58].

In contrast, supportive technological systems for mobility already count on a considerably high (and still increasing) acceptance in countries from other regions, principally Europe and Asia. In these contexts, a strong satisfaction with these systems is usually reported, as a good part of the population reports perceiving everyday mobility-related and quality of life improvements as a result of these advances [59]. Despite the evident contextual and income-related differences, it can be hypothesized that increasing community involvement, participation, and individual responsiveness for the correct functioning of ITS systems might increase their acceptance, valuation, and future use [60].

In this regard, differences in the degree of knowledge of ITS systems have been stated as closely linked to the digital (income-based) gap between territories [61]. While in developed countries, the majority of people have access to electronic devices, digital media and the internet, low and middle-income countries (LMICs) still tend to report a very low digitization rate. According to the data, only between 6–7 out of 10 Dominicans have an internet connection [62], placing them very far from high-income countries (HICs), where this index usually reaches 90% of people [63]. Notwithstanding, and more as a potentially positive scenario, the Digital Agenda 2030 for the Dominican Republic has been recently approved with the aim of promoting a digital, open, and participatory society in which connectivity, cybersecurity, and digital accessibility of the population could be improved in several fields, including transport and mobility [27,64].

4.2. Present and Future: Traffic Management and Environmental Sustainability

Apart from being considerably favourable, the valuation of new technologies applied to mobility, ITSs, and mobile apps has also been significantly associated with having a greater concern about environmental issues [65]. In other words, the people considering environmental issues as relevant problems for the country were the ones who best perceive the benefits that digital tools can bring to the traffic environment.

Indeed, the reality is that the Dominican Republic has concerningly high rates of pollution, and this can be an attractive opportunity to change this challenging state of affairs [66]. In recent years, the renovation of vehicles to reduce emissions has been boosted. However, currently, more than 52% of the country's vehicle fleet is composed of vehicles older than 10 years, which produce excess polluting gases [67]. This situation, together with the dynamics of Dominican commuting, does not favour the reduction of emissions produced by transportation. A large part of the population travels by private vehicle, with very few people opting for bicycles or walking for their regular trips [38,50,51]. Therefore, ITS systems that manage and regulate traffic and public transport could be very beneficial for the environment. Their implementation would reduce traffic jams in urban areas and minimize the time spent commuting, with a direct impact on the country's pollution rates [68]. To complement this measure, it would be highly recommended to improve the conditions of sidewalks and pedestrian areas to encourage travel by non-polluting means of transport.

Transport dynamics can also explain the relationships between other variables. Thus, the degree of public transport use is related to the perception of how much a traffic application can contribute to mobility. At the same time, the greater use of private transport is also related to the prioritization of ITS systems. Thus, frequent drivers consider the implementation of certain ITS services to be more urgent.

4.3. Perceived Value and Usage of Mobile Apps in Mobility

Although the use of applications while travelling is relatively low, it has grown substantially in the short, analyzed period of time. Therefore, it is expected that this positive trend will continue in the coming years, especially with the increase in the acquisition of mobile devices by the population.

The main applications used are related to maps and route settings. There is, in any case, a small percentage using active transport-related apps, especially those useful to strengthen walking and cycling trips, something that could be used in favor of public health improvements in the mid- and long-term, even though their distracting potential should also be considered [69–71].

Additionally, and significantly linked to other potentially helpful technological advances for enhancing urban mobility, the role of emerging trends in transportation should be considered, such as the globally growing use of micromobility [72]. Although its market share is still minimal (estimated to be <4%), partly as a result of the low-income status of the country and the lack of infrastructural adequation for it, the potential improvements that these trends could bring for enhancing transport dynamics in the region (as well as their 'side effects') remain to be seen.

Changes in citizens' travel habits should be promoted with concrete measures and initiatives ranging from reducing the prices to improving infrastructure and public vehicles. Once these adjustments have been made, it is more likely that citizens will be able to use the application to the fullest extent of its functionalities. In this way, the applications and other technological factors can act as enablers of change in the transportation dynamics of Dominicans.

5. Conclusions

This study shows how there is a relatively high acceptance, attributed value, and attitudinal predisposition towards both intelligent transportation systems (ITS) and various support technologies applicable to mobility. However, their actual usage rates remain considerably low, probably enhanced by the low and middle-income status of the country.

Further, these findings suggest the need to strengthen information and communication flows over emerging technologies, as several gaps remain present, even though a slight increase in knowledge, acceptance, and engagement on mobility-related applications and other technologies has been observed between the years 2018 and 2019.

In practical settings, this research can contribute to understanding technology-related perceptions, beliefs, and practices and their potential applications for traffic, mobility, and road safety. This will be particularly relevant to increasing users' awareness, knowledge, and confidence on the matter, favoring their potential application to improve transportation dynamics, quality of life, and traffic safety in the country.

Author Contributions: S.A.U., F.A. and M.F. conceived and designed the research, and performed the data collection; M.F. analyzed the data; F.A., M.T.T. and S.A.U. contributed with reagents/materials/analysis tools; M.T.T. contributed with paper revisions; M.F. and S.A.U. wrote and revised the paper. All authors have read and agreed to the published version of the manuscript.

Funding: This study was funded by the National Institute of Transit and Land Transportation (INTRANT) and its Permanent Observatory in Road Safety (OPSEVI; public agency of the Dominican Republic)-Grant number: 20170475. Additionally, this work was supported by the research grant ACIF/2020/035 (MF) from "Generalitat Valenciana". Funding entities did not contribute to the study design or data collection, analysis and interpretation, or writing of the manuscript.

Institutional Review Board Statement: Both data collection phases of this study were conducted according to the guidelines provided by the Declaration of Helsinki and approved by the Ethics Committee of Research in Social Sciences in Health" from the Institute on Traffic and Road Safety at the University of Valencia, Spain (IRB numbers HE0001251018—25 October 2018, and HE0002130519—5 May 2019).

Informed Consent Statement: Informed consent was obtained from all subjects involved in the study.

Data Availability Statement: The data will be available upon reasonable request to the corresponding author.

Acknowledgments: The authors wish to thank Arash Javadinejad for the professional edition of the final version of the manuscript.

References

1. Dimitrakopoulos, G.J.; Uden, L.; Varlamis, I. *The Future of Intelligent Transport Systems*; Elsevier: Amsterdam, The Netherlands, 2020.
2. Perallos, A.; Hernandez-Jayo, U.; Onieva, E.; Zuazola, I.J.G. *Intelligent Transport Systems: Technologies and Applications*; John Wiley & Sons: Hoboken, NJ, USA, 2015.
3. Zhu, L.; Yu, F.R.; Wang, Y.; Ning, B.; Tang, T. Big data analytics in intelligent transportation systems: A survey. *IEEE Trans. Intell. Transp. Syst.* **2018**, *20*, 383–398. [CrossRef]
4. Docherty, I.; Marsden, G.; Anable, J. The governance of smart mobility. *Transp. Res. Part A Policy Pract.* **2018**, *115*, 114–125. [CrossRef]
5. Mangiaracina, R.; Perego, A.; Salvadori, G.; Tumino, A. A comprehensive view of intelligent transport systems for urban smart mobility. *Int. J. Logist. Res. Appl.* **2017**, *20*, 39–52. [CrossRef]
6. Greenblatt, J.B.; Shaheen, S. Automated Vehicles, On-Demand Mobility, and Environmental Impacts. *Curr. Sustain. Renew. Energy Rep.* **2015**, *2*, 74–81. [CrossRef]
7. Teslyuk, V.; Beregovskyi, V.; Denysyuk, P.; Teslyuk, T.; Lozynskyi, A. Development and implementation of the technical accident prevention subsystem for the smart home system. *Int. J. Intell. Syst. Appl.* **2018**, *12*, 1. [CrossRef]
8. Scholliers, J.; Van Sambeek, M.; Moerman, K. Integration of vulnerable road users in cooperative ITS systems. *Eur. Transp. Res. Rev.* **2017**, *9*, 15. [CrossRef]
9. Faria, R.; Brito, L.; Baras, K.; Silva, J. Smart mobility: A survey. In Proceedings of the 2017 International Conference on Internet of Things for the Global Community (IoTGC), Funchal, Portugal, 10–13 July 2017; pp. 1–8. [CrossRef]
10. Thomopoulos, N.; Givoni, M. The autonomous car—A blessing or a curse for the future of low carbon mobility? An exploration of likely vs. desirable outcomes. *Eur. J. Futures Res.* **2015**, *3*, 14. [CrossRef]
11. Albino, V.; Berardi, U.; Dangelico, R.M. Smart cities: Definitions, dimensions, performance, and initiatives. *J. Urban Technol.* **2015**, *22*, 3–21. [CrossRef]
12. Castellanos, S.; Grant-Muller, S.; Wright, K. Technology, transport, and the sharing economy: Towards a working taxonomy for shared mobility. *Transp. Rev.* **2022**, *42*, 318–336. [CrossRef]
13. Eskandarian, A.; Wu, C.; Sun, C. Research advances and challenges of autonomous and connected ground vehicles. *IEEE Trans. Intell. Transp. Syst.* **2019**, *22*, 683–711. [CrossRef]
14. Pendleton, S.D.; Andersen, H.; Du, X.; Shen, X.; Meghjani, M.; Eng, Y.H.; Rus, D.; Ang, M.H. Perception, planning, control, and coordination for autonomous vehicles. *Machines* **2017**, *5*, 6. [CrossRef]
15. Sun, L.; Zhan, W.; Chan, C.Y.; Tomizuka, M. Behavior planning of autonomous cars with social perception. In Proceedings of the 2019 IEEE Intelligent Vehicles Symposium (IV), Paris, France, 9–12 June 2019; pp. 207–213. [CrossRef]
16. Cordts, P.; Cotten, S.R.; Qu, T.; Bush, T. Mobility challenges and perceptions of autonomous vehicles for individuals with physical disabilities. *Disabil. Health J.* **2021**, *14*, 101131. [CrossRef] [PubMed]
17. Hulse, L.M.; Xie, H.; Galea, E.R. Perceptions of autonomous vehicles: Relationships with road users, risk, gender and age. *Saf. Sci.* **2018**, *102*, 1–13. [CrossRef]
18. Bansal, P.; Kockelman, K.M.; Singh, A. Assessing public opinions of and interest in new vehicle technologies: An Austin perspective. *Transp. Res. Part C Emerg. Technol.* **2016**, *67*, 1–14. [CrossRef]
19. Mushtaq, A.; Riaz, S.; Mohd, H.; Saleh, A. Perception and technology adoption trends for autonomous vehicles: Educational case study. In Proceedings of the 2018 Advances in Science and Engineering Technology International Conferences (ASET), Dubai, Sharjah, Abu Dhabi, United Arab Emirates, 6 February–5 April 2018; pp. 1–5. [CrossRef]
20. Kassens-Noor, E.; Cai, M.; Kotval-Karamchandani, Z.; Decaminada, T. Autonomous vehicles and mobility for people with special needs. *Transp. Res. Part A Policy Pract.* **2021**, *150*, 385–397. [CrossRef]
21. König, M.; Neumayr, L. Users' resistance towards radical innovations: The case of the self-driving car. *Transp. Res. Part F Traffic Psychol. Behav.* **2017**, *44*, 42–52. [CrossRef]
22. Begg, D. A 2050 Vision for London: What Are the Implications of Driverless Transport? 2014. Available online: https://trid.trb.org/view/1319762 (accessed on 27 December 2021).
23. Casley, S.V.; Quartulli, A.M.; Jardim, A.S. A Study of Public Acceptance of Autonomous Cars. 2013. Available online: https://digital.wpi.edu/concern/student_works/pz50gw37k (accessed on 27 December 2021).
24. Gkartzonikas, C.; Gkritza, K. What have we learned? A review of stated preference and choice studies on autonomous vehicles. *Transp. Res. Part C Emerg. Technol.* **2019**, *98*, 323–337. [CrossRef]
25. Xu, X.; Fan, C.K. Autonomous vehicles, risk perceptions and insurance demand: An individual survey in China. *Transp. Res. Part A Policy Pract.* **2019**, *124*, 549–556. [CrossRef]

26. Menon, N.; Pinjari, A.; Zhang, Y.; Zou, L. Consumer Perception and Intended Adoption of Autonomous-Vehicle Technology: Findings from a University Population Survey (No. 16-5998). 2016. Available online: https://trid.trb.org/view/1394249 (accessed on 28 December 2021).

27. Alonso, F.; Faus, M.; Esteban, C.; Useche, S.A. Is There a Predisposition towards the Use of New Technologies within the Traffic Field of Emerging Countries? The Case of the Dominican Republic. *Electronics* **2021**, *10*, 1208. [CrossRef]

28. Gray, T.J.; Gainous, J.; Wagner, K.M. Gender and the digital divide in Latin America. *Soc. Sci. Q.* **2017**, *98*, 326–340. [CrossRef]

29. Nadhom, M.; Loskot, P. Survey of public data sources on the Internet usage and other Internet statistics. *Data Brief* **2018**, *18*, 1914–1929. [CrossRef] [PubMed]

30. Ackaah, W.; Osei, K.K. Perception of autonomous vehicles—A Ghanaian perspective. *Transp. Res. Interdiscip. Perspect.* **2021**, *11*, 100437. [CrossRef]

31. Brell, T.; Philipsen, R.; Ziefle, M. sCARy! Risk perceptions in autonomous driving: The influence of experience on perceived benefits and barriers. *Risk Anal.* **2019**, *39*, 342–357. [CrossRef] [PubMed]

32. Schoettle, B.; Sivak, M. A Survey of Public Opinion about Autonomous and Self-Driving Vehicles in the US, the UK, and Australia. University of Michigan, Ann Arbor, Transportation Research Institute. 2014. Available online: https://hdl.handle.net/2027.42/108384 (accessed on 25 December 2021).

33. Shabanpour, R.; Golshani, N.; Shamshiripour, A.; Mohammadian, A.K. Eliciting preferences for adoption of fully automated vehicles using best-worst analysis. *Transp. Res. Part C Emerg. Technol.* **2018**, *93*, 463–478. [CrossRef]

34. Liljamo, T.; Liimatainen, H.; Pöllänen, M. Attitudes and concerns on automated vehicles. *Transp. Res. Part F Traffic Psychol. Behav.* **2018**, *59*, 24–44. [CrossRef]

35. Haboucha, C.J.; Ishaq, R.; Shiftan, Y. User preferences regarding autonomous vehicles. *Transp. Res. Part C Emerg. Technol.* **2017**, *78*, 37–49. [CrossRef]

36. Schifter, D.E.; Ajzen, I. Intention, perceived control, and weight loss: An application of the theory of planned behavior. *J. Pers. Soc. Psychol.* **1985**, *49*, 843. [CrossRef]

37. Davis, F.D. Utilidad percibida, facilidad de uso percibida y aceptación de las tecnologías de la información por parte del usuario. *MIS Q.* **1989**, *13*, 21. [CrossRef]

38. Alonso, F.; Useche, S.A.; Faus, M.; Esteban, C. Does urban security modulate transportation choices and travel behavior of citizens? A national study in the Dominican Republic. *Front. Sustain. Cities* **2020**, *2*, 42. [CrossRef]

39. Alonso, F.; Faus, M.; Cendales, B.; Useche, S.A. Citizens' Perceptions in Relation to Transport Systems and Infrastructures: A Nationwide Study in the Dominican Republic. *Infrastructures* **2021**, *6*, 153. [CrossRef]

40. Nazif-Munoz, J.I.; Batomen, B.; Nandi, A. Does ridesharing affect road safety? The introduction of Moto-Uber and other factors in the Dominican Republic. *Res. Glob.* **2021**, *4*, 100077. [CrossRef]

41. Alonso, F.; Faus, M.; Fernández, C.; Useche, S.A. "Where Have I Heard It?" Assessing the Recall of Traffic Safety Campaigns in the Dominican Republic. *Energies* **2021**, *14*, 5792. [CrossRef]

42. Nazif, J.I.; Pérez, G. Revisión del Desempeño de la Seguridad vial en la República Dominicana. 2018. Available online: http://hdl.handle.net/11362/44211 (accessed on 28 December 2021).

43. Rodríguez, D.A.; Santana, M.; Pardo, C.F. *La Motocicleta en América Latina: Caracterización de su Uso e Impactos en la Movilidad en Cinco Ciudades de la Región*; CAF: Washington, DC, USA, 2015. Available online: http://scioteca.caf.com/handle/123456789/754 (accessed on 26 December 2021).

44. Hernández, B.B.; Osorio, P.F.C. El servicio de mototaxis: Una fuente alternativa de trabajo en Puebla. *DÍKÊ Rev. Investig. Derecho Criminol. Consult. Jurídica* **2017**, *8*, 151–171. [CrossRef]

45. Santana, E.A.; Marte, J.C. Propuesta Estratégica para la Mejora en la Calidad del Servicio de Transporte Público: Caso Transporte Expreso Tarea, Ruta Santo Domingo-Bonao. República Dominicana. 2017. Available online: http://investigare.pucmm.edu.do:8080/xmlui/handle/20.500.12060/1861 (accessed on 21 December 2021).

46. Useche, S.A.; Montoro, L.; Alonso, F.; Tortosa, F. Does gender really matter? A structural equation model to explain risky and positive cycling behaviors. *Accid. Anal. Prev.* **2018**, *118*, 86–95. [CrossRef] [PubMed]

47. Sunio, V.; Joseph Li, W.; Pontawe, J.; Dizon, A.; Bienne Valderrama, J.; Robang, A. Service contracting as a policy response for public transport recovery during the Covid-19 Pandemic: A preliminary evaluation. *Transp. Res. Interdiscip. Perspect.* **2022**, *13*, 100559. [CrossRef] [PubMed]

48. Hasan, U.; Whyte, A.; Al Jassmi, H. A review of the transformation of road transport systems: Are we ready for the next step in artificially intelligent sustainable Transport? *Appl. Syst. Innov.* **2019**, *3*, 1. [CrossRef]

49. Montañez, M.R. Un nuevo modelo de transporte para el gran Santo Domingo. *Cienc. Soc.* **2016**, *41*, 337–359. [CrossRef]

50. INTRANT. *National Mobility Survey of the Dominican Republic, Results REPORT 2018*; Instituto Nacional de Tránsito y Transporte Terrestre: Santo Domingo, Dominican Republic, 2019.

51. INTRANT. *National Mobility Survey of the Dominican Republic, Results REPORT 2019*; Instituto Nacional de Tránsito y Transporte Terrestre: Santo Domingo, Dominican Republic, 2020.

52. Trager, J.; Kalová, L.; Pagany, R.; Dorner, W. Warning apps for road safety: A technological and economical perspective for autonomous driving–the warning task in the transition from human driver to automated driving. *Int. J. Hum.-Comput. Interact.* **2021**, *37*, 363–377. [CrossRef]

53. Hou, J.J.; Arpan, L.; Wu, Y.; Feiock, R.; Ozguven, E.; Arghandeh, R. The road toward smart cities: A study of citizens' acceptance of mobile applications for city services. *Energies* **2020**, *13*, 2496. [CrossRef]
54. Tintin, R.A.; Vela, M.; Anzules, V.; Escobar, V. Smart cities and telecommuting in Ecuador. In Proceedings of the 2015 Second International Conference on eDemocracy & eGovernment (ICEDEG), Quito, Ecuador, 8–10 April 2015; pp. 49–53. [CrossRef]
55. Rodríguez Martínez, J.A. Analisis de la implementacion de Smart Cities en Colombia, caso Bogotá y Medellín 2016–2019. Available online: http://hdl.handle.net/10554/53940 (accessed on 26 December 2021).
56. Rojas, O.A.V. Adaptación ciudadana a las Tecnologías de Información y Comunicación en "Smart Cities" desde una perspectiva de la educación para el desarrollo sostenible, caso Medellín-Colombia. *Rev. Mex. Cienc. Agrícolas* **2015**, *1*, 487–494. Available online: https://www.redalyc.org/pdf/2631/263139243066.pdf (accessed on 31 December 2021).
57. Negro, A.E. La promesa de las "smart cities" como nuevo enclave ideológico del proceso de neoliberalización de las ciudades. *Quid 16 Rev. Área Estud. Urbanos* **2021**, 244–262. Available online: https://dialnet.unirioja.es/servlet/articulo?codigo=8139864 (accessed on 24 December 2021).
58. Schreiner, C. International case studies of smart cities: Rio de Janeiro, Brazil. *Inter-Am. Dev. Bank* **2016**. Available online: https://publications.iadb.org/en/international-case-studies-smart-cities-rio-de-janeiro-brazil (accessed on 24 December 2021).
59. Macke, J.; Casagrande, R.M.; Sarate, J.A.R.; Silva, K.A. Smart city and quality of life: Citizens' perception in a Brazilian case study. *J. Clean. Prod.* **2018**, *182*, 717–726. [CrossRef]
60. Lytras, M.D.; Visvizi, A.; Chopdar, P.K.; Sarirete, A.; Alhalabi, W. Information Management in Smart Cities: Turning end users' views into multi-item scale development, validation, and policymaking recommendations. *Int. J. Inf. Manag.* **2021**, *56*, 102146. [CrossRef]
61. Naji, J.E. Educación en medios ante la brecha digital en los países del Sur. *Comun. Rev. Científica Comun. Educ.* **2009**, *16*, 41–50. [CrossRef]
62. Del Carmen, G.; Díaz, K.; Ruiz-Arranz, M. A un clic de la Transición: Economía Digital en Centroamérica y la República Dominicana. 2020. Available online: https://siip.produccion.gob.bo/noticias/files/2020-cde0b-3clic.pdf (accessed on 10 February 2022).
63. Instituto Nacional de Estadística (INE). Población que Usa Internet (en los Últimos tres Meses). Tipo de Actividades Realizadas por Internet. 2020. Available online: https://www.ine.es/ss/Satellite?L=es_ES&c=INESeccion_C&cid=1259925528782&p=1254735110672&pagename=ProductosYServicios%2FPYSLayout (accessed on 27 December 2021).
64. Gobierno de la República Dominicana Agenda Digital 2021. República Dominicana. 2021. Available online: https://agendadigital.gob.do/ (accessed on 27 December 2021).
65. Nowland, R.; Necka, E.A.; Cacioppo, J.T. Loneliness and social internet use: Pathways to reconnection in a digital world? *Perspect. Psychol. Sci.* **2018**, *13*, 70–87. [CrossRef]
66. Ratick, S.J.; Osleeb, J.P. Measuring the vulnerability of populations susceptible to lead contamination in the Dominican Republic: Evaluating composite index construction methods. *GeoJournal* **2013**, *78*, 259–272. [CrossRef]
67. Spomenka Angelov MS, I.E.; Angelov, A. Parque vehicular de la República Dominicana en el futuro uso de tecnologías limpias. In Proceedings of the 15th LACCEI International Multi-Conference for Engineering, Education, and Technology: "Global Partnership for Development and Engineering Education", Boca Raton, FL, USA, 19–21 July 2017. Available online: http://www.laccei.org/LACCEI2017-BocaRaton/work_in_progress/WP508.pdf (accessed on 30 December 2021).
68. Talari, S.; Shafie-Khah, M.; Siano, P.; Loia, V.; Tommasetti, A.; Catalão, J.P. A review of smart cities based on the internet of things concept. *Energies* **2017**, *10*, 421. [CrossRef]
69. Hudák, M.; Madleňák, R. The research of driver distraction by visual smog on selected road stretch in Slovakia. *Procedia Eng.* **2017**, *178*, 472–479. [CrossRef]
70. Horsman, G.; Conniss, L.R. Investigating evidence of mobile phone usage by drivers in road traffic accidents. *Digit. Investig.* **2015**, *12*, S30–S37. [CrossRef]
71. INTRANT. *APP Sistema Integrado de Transporte Público del Gran Santo Domingo*; Instituto Nacional de Tránsito y Transporte Terrestre: Santo Domingo, Dominican Republic, 2019.
72. Orozco-Fontalvo, M.; Llerena, L.; Cantillo, V. Dockless electric scooters: A review of a growing micromobility mode. *Int. J. Sustain. Transp.* **2022**, 1–17. [CrossRef]

Article

Influence of Mix Proportions on Rheological Properties, Air Content of Wet Shotcrete—A Case Study

Jun Xie [1,2,3]**, Xiangfei Cui** [1,2,]*****, **Nan Guo** [1,3] **and Guoming Liu** [1,2,3,]*****

[1]   College of Energy and Mining Engineering, Shandong University of Science and Technology, Qingdao 266590, China; xiejunsdust@sdust.edu.cn (J.X.); felixwzshi@gmail.com (N.G.)
[2]   State Key Laboratory of Strata Intelligent Control and Green Mining Co-founded by Shandong Province and the Ministry of Science and Technology, Qingdao 266590, China
[3]   College of Safety and Environmental Engineering, Shandong University of Science and Technology, Qingdao 266590, China
*****   Correspondence: skd994394@sdust.edu.cn (X.C.); skd995978@sdust.edu.cn (G.L.)

Abstract: To study the influence of different mix proportions on the fresh properties of wet shotcrete, the rheological properties and air content of wet shotcrete with different admixtures before pumping were measured. In addition, the pressure drop along the pipeline and the build-up thickness were studied, and the relationship between the rheological properties and the pumpability and sprayability was discussed. This paper attempts to reveal the influence mechanism of admixtures on the fluidity of wet shotcrete by means of pictures. The results show that free paste effect and ball effect are two key factors that affect the performance of fresh wet shotcrete. Air-entraining agent and fly ash are commonly used admixtures, which improve the pumping performance and spraying performance. Finally, the mix proportions of wet shotcrete are put forward to meet the requirements of different types of shotcrete.

Keywords: wet shotcrete; mix proportion; fresh properties; air content

Citation: Xie, J.; Cui, X.; Guo, N.; Liu, G. Influence of Mix Proportions on Rheological Properties, Air Content of Wet Shotcrete—A Case Study. *Appl. Sci.* **2021**, *11*, 3550. https://doi.org/10.3390/app11083550

Academic Editor: José A. Orosa

Received: 13 March 2021
Accepted: 13 April 2021
Published: 15 April 2021

Publisher's Note: MDPI stays neutral with regard to jurisdictional claims in published maps and institutional affiliations.

1. Introduction

Wet shotcrete is a kind of self-compacting concrete technology, in which fresh wet shotcrete is pumped into the pipeline, and then sprayed onto the receiving surface under the function of compressed air [1–7]. Wet shotcrete is widely used in the construction industry, such as slope maintenance and underground space engineering. Especially in the complex coal mine roadway, compared with dry mix shotcrete, wet shotcrete has the advantages of low dust concentration and less rebound, which reduces the mining hazards such as pneumoconiosis.

In the application of wet shotcrete, the pumping process is the initial important process to ensure successful conveying without blockage. Pumpability refers to the pumping capacity of fresh wet shotcrete, i.e., fresh concrete should have proper fluidity and stability when transported through pipes under pressure. At present, a lot of research has been devoted to the study of measurement techniques that evaluate the pumpability of fresh concrete. Especially in the field of self-compacting concrete, rheological and tribological properties of fresh concrete have been recognized as two key factors that affect the pumping performance of fresh concrete [7,8].

Fresh concrete ingredients affect the rheological properties. Literature [8] reviewed the effects of internal components of fresh concrete and external admixtures on rheological properties, including cement, fly ash, slag, silica fume, limestone powder, coarse and fine aggregates, and admixtures (water-reducing agent, viscosity agent, and air-entraining agent). Dils et al. [9,10] studied the effect of cement fineness and chemical composition on the rheological properties of concrete, and pointed out that the larger the surface area of the cement, the higher the content of tricalcium aluminate and alkali, the worse the working performance of fresh concrete. Koehler et al. [11] found that the increase of cement paste content reduced the plastic viscosity and yield stress of fresh concrete; especially for mixtures with large differences in aggregate shapes and poor particle gradation, and increasing the cement paste content has a positive effect on concrete pumping. Choi et al. [12] studied the effect of different coarse aggregate sizes on the rheological properties of fresh concrete, as the aggregate size increases, the plastic viscosity increased. Xu et al. [13] studied the rheological and mechanical property of fresh cemented materials incorporating silica fume, and assessed the applicability of using silica fume as a partial cement replacement. Park et al. [14] obtained through experiments: the rheological parameters (plastic viscosity and shear stress) of fresh concrete mixed with fly ash are lower than those of concrete without fly ash. In addition, with the increase of fly ash content increases, the plastic viscosity and shear stress gradually increase. The addition of silica fume increases the plastic viscosity and yield stress of the fresh concrete, and enhances the flocculation inside the concrete [15,16]. Yun et al. [17,18] used the IBB rheometer to measure the influence of mineral admixtures on the performance of wet shotcrete, and concluded that the pumpability and shotcretability of fresh concrete can be improved by adding silica fume and fly ash at the same time.

In the early stage, slump and pressure bleeding tests were combined to predict the pumpability of fresh concrete [19,20]. However, the good slump cannot guarantee good pumping performance in field application. Burns [21] studied the pumping pressure according to the rheological and tribological characteristics of fresh wet shotcrete moving in small diameter hose. Based on the full-scale pumping test, Feys et al. [22] analyzed the influence of fresh concrete rheology, tribology, pipe radius, and flow rate on pumping pressure. Gołaszewski [23] pointed out that according to the test results of BT2 rheometer, the change of rheological properties of fresh concrete with time could be estimated based on the rheological properties of corresponding mortar. Concrete flow in pipes can be summarized as a combination of three flow types, including lubrication layer shear, shear flow and plug flow [24–27]. Secrieru et al. [19,28] indicated that the pumping performance of fresh concrete depends on the formation of the lubricating layer and the rheological properties of concrete plug. Mixing various chemical admixtures or mineral additives can have a significant impact on the rheological properties of concrete, thus affecting the pumping performance, at the same time, improving the stability of the mixture, and even changing the thickness of the lubrication layer [28,29]. However, the differences in the composition of mineral additives or chemical admixtures lead to different rheological properties. According to the previous experiments and literature, the pumping performance and sprayability can be estimated by rheological parameters. For example, higher flow resistance (yield shear stress) means better sprayability and poorer pumpability.

To study the rheological properties of wet shotcrete with different admixtures and further estimate the pumpability and shotcreting performance before pumping, this paper studies the influence of water–binder ratio (w/b, WC), sand ratio (SR), fiber length (FI), air-entraining agent (AE), water-reducing agent (WR), silica fume (SF) and fly ash (FA) on the rheological properties and air content of wet shotcrete. The influence mechanism of admixtures on the properties of fresh concrete is discussed. By comparing the effect of fly ash, silica fume, and other admixtures, the application of wet shotcrete is improved. It is hoped that this study may provide a useful information for improving the pumping and spraying properties of fresh wet shotcrete and further improving the quality of wet shotcrete. Through this research, the main goal is to provide a simplified guide for shotcrete technologists.

2. Materials and Methods

2.1. Materials

Table 1 shows the mix proportions. Cement is the ordinary Portland cement (OPC) Po. 42.5, specific gravity 3.14. The F grade fly ash with specific gravity of 2.34 is selected in this test. These two chemical compositions are shown in Table 2. The particle size distribution of cement, fly ash and silica fume was reported in [30]. Please note that these mix proportions are based on previous tests and published literature [2,3,31–36]. Due to the different characteristics of raw materials from different sources, the mix proportion is adjusted according to ACI 506R-16 of "Shotcrete Guidelines" [37]. For example, the selected water–binder ratio is slightly higher.

Table 1. Mixture proportions of fresh wet shotcrete.

No.	Variable	WC (w/b)	Water (kg/m^3)	SR(%)	Sand (kg/m^3)	FI (mm)	AEA (%)	WR (%)	SF (%)	FA (%)
WC45	Water–binder ratio (WC)	0.45	198	70	1100			0.3		
WC50		0.5	220	68	1080					
WC55		0.55	242	66	1070					
WC60		0.6	264	66	1050					
SR50	Sand ratio (SR)	0.5	220	50	900					
SR60		0.5	220	60	1000					
SR70		0.5	220	70	1100			0.3		
FI6	Fiber (FI)	0.5	220	60	1000	6		0.3		
FI12		0.5	220	60	1000	12		0.3		
FI29		0.5	220	60	1000	29		0.3		
AE0.02%	Air-entraining agent (AEA)	0.5	220	60	1000		0.02			
AE0.04%		0.5	220	60	1000		0.04			
AE0.06%		0.5	220	60	1000		0.06			
WR0.3%	Water reducer (WR)	0.5	220	60	1000			0.3		
WR0.6%		0.5	220	60	1000			0.6		
WR0.9%		0.5	220	60	1000			0.9		
WR1.2%		0.5	220	60	1000			1.2		
SF5%	Silica fume (SF)	0.5	220	60	1000			0.3	5	
SF10%		0.5	220	60	1000			0.3	10	
SF15%		0.5	220	60	1000			0.3	15	
FA5%	Fly ash (FA)	0.5	220	60	1000					5
FA10%		0.5	220	60	1000					10
FA15%		0.5	220	60	1000					15

Table 2. Chemical compositions of cement and fly ash (unit: %).

Raw Materials	SiO_2	Al_2O_3	Fe_2O_3	CaO	MgO	SO_3
Cement	19.5	6.45	3.08	57.57	1.21	2.01
Fly ash	43.64	25.39	4.19	5.62	0.84	0.28
Silica fume	95	1.1	0.8	0.3	0.6	-

In this study, silica fume with specific gravity of 2.21 and specific surface area of 160,000–300,000 cm^2/g were used. The silica fume is composed of up to 95% SiO_2 and less than 1% CaO, as listed in Table 2. In this study, natural river sand was used as fine aggregate with fineness modulus of 2.66. For coarse aggregate, gravel with a maximum size of 10 mm is used, the fineness modulus is 5.70. The specific gravity of sand and gravel are 2.61 and 2.67, respectively. At the same time, the water absorption of fine aggregate and coarse aggregate is considered to modify the mixed water. The aggregate grading curve conforms to the national standard GB50086-2001. In this study, polypropylene fiber (FI) with specific gravity of 0.93 was used. The fiber diameter range was 18–48 µm, and the fiber content in shotcrete was 0.9 kg/m^3. The tensile strength and elastic modulus of the fiber are more than 500 MPa and 3850 MPa, respectively. Polycarboxylate water reducer (WR) and 126A type air-entraining agent (AEA) were mixed into wet shotcrete respectively. The water-reducing rate of the water reducer is about 24%. Admixtures are marked as %, indicating the percentage of the admixture relative to the binder content (mass). Each fresh concrete sample was produced in a horizontal axis forced mixer. The mixing process is as follows: first mixing coarse aggregate and sand, then adding cement, fly ash, fiber and silica fume in 20 s, and then adding water and other additives in 2 min.

2.2. Measurement of Rheological Properties of Wet Shotcrete

The rheological properties of fresh wet shotcrete were tested by eBT2 rheometer. This rheometer is different from the coaxial cylinder rational viscometer. Figure 1a shows the structure of eBT2, including a shaft decoder and two moment sensitive probes. The smart phone controls the rheometer through Bluetooth interface. It has been confirmed that the behavior of fresh concrete is similar to Bingham fluid (Equation (1)), where τ (Pa) is the shear stress, τ_0 (Pa) is the yield stress, and η (Pa·s) is the plastic viscosity [38–41]. The corresponding relationship can be expressed by Equation (2), where g (Nm) and H (Ns) are flow resistance and torque viscosity respectively, which are called relative yield stress and relative plastic viscosity [42,43]. The rheometer can determine the flow resistance and torque viscosity according to the measurement of momentum and rotation speed. Due to the double probe structure and the special operation in the measurement process, the problems of segregation and structural breakdown in the measurement process are avoided. The rules and methods of using this rheometer were showed in detail in [44–47]. The rotational speed setting during the rheological test is shown in Figure 1b. One rotation time is 30 s.

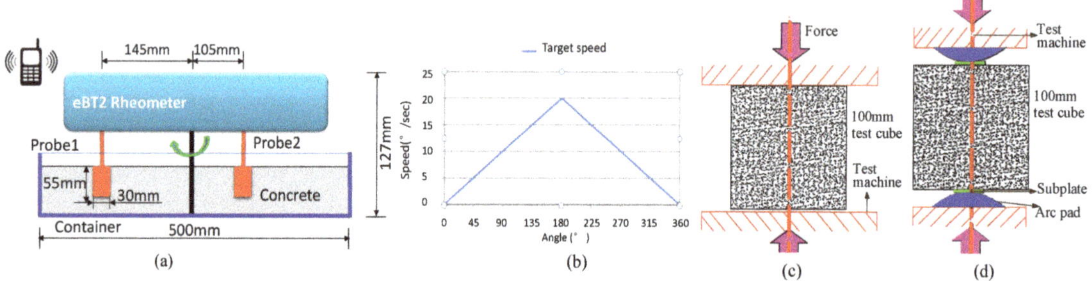

Figure 1. Schematic diagram of eBT2 rheometer (**a**), rotational speed (**b**), compressive test (**c**) and splitting test (**d**).

$$\tau = \tau_0 + \eta\,\dot{\gamma} \tag{1}$$

$$T = g + h\,N \tag{2}$$

2.3. Air Content and Slump Test

The air content of fresh concrete was measured by LA-0316 type air content meter. The air content measuring instrument is composed of base, vent valve, pressure chamber, and pressure gauge. The air content value can be read directly from the pressure gauge. Slump value was measured by slump cone. Two tests were carried out according to the requirements of Chinese standard GB/T50086-2016. Measurement method of air content: (1) Fill base with concrete and strike off. Clean cover and base rim; clamp together. (2) Open petcocks. Use syringe to inject water through one petcock until water is expelled from opposite petcock. (3) Close air bleeder value on air chamber. Pump up air initial pressure line gauge. Wait a few seconds, tap gauge lightly. If necessary, add or release air to attain reading at initial pressure line. Close petcocks. (4) Press needle valve lever to release air into base. Continue pressing lever and lightly tap gauge. Read direct percentage of air. (5) Thoroughly clean base, cover, and petcock openings with running water. Measurement method of slump: The slump is measured by the slump cone, which is a 300 mm high trumpet shaped drum. Pour concrete into the slump cone and fill it in three times. After each filling, use a tamping hammer to strike 25 times from the outside to the inside along the wall of the cone. After tamping, level it. Then pull up the slump cone vertically, and the fresh concrete will collapse due to its own weight. The slump value is obtained by subtracting the height of the cone from the height of the highest point of the collapsed concrete.

2.4. Measurement of Hardenability

To evaluate the mechanical properties of wet shotcrete before pumping, wet shotcrete is directly poured into concrete mold. According to the Chinese Standard "Test method for fiber reinforced concrete" (CECS13:2009) and "Concrete quality control standard" (GB 50164-2011), the curing and mechanical tests of specimens were carried out. After 1 day curing in a standard curing chamber with temperature of 20 ± 2 °C and 95% relative humidity, specimens were demolded. Then specimens were cured at standard condition to 28 days. Those specimens were prepared for the measurement of mechanical properties of shotcrete, the methodology is shown in Figure 1c,d. The 28 d compressive strength varied between 20.3 (FI29) and 42.6MPa (SF15), while the 28 d splitting strength varied between 3.6 (SR70) and 5.2 MPa (FI12).

2.5. Pumpability and Shootability Tests

Based on the pumping and spraying tests of fresh concrete, the relationship between rheological properties and workability (pumpability and injectability) was studied. SPB7 wet spraying machine was used in the test. The length of the pipe is 25 m, the wind pressure is 0.6 MPa, the spray distance is 1 m, and the spray angle is 70–90 degrees. Figure 2 shows the experimental layout. The pressure sensor was installed in different positions. The pressure drop represents pumpability, and the build-up thickness represents shootability [32,48]. The pressure drop (ΔP) was calculated from the measured pressure value. The measurement way of build-thickness after shotcreting is shown in Figure 2. Lower pressure drop means better pumping performance; thicker build-up thickness means better shootability, and vice versa.

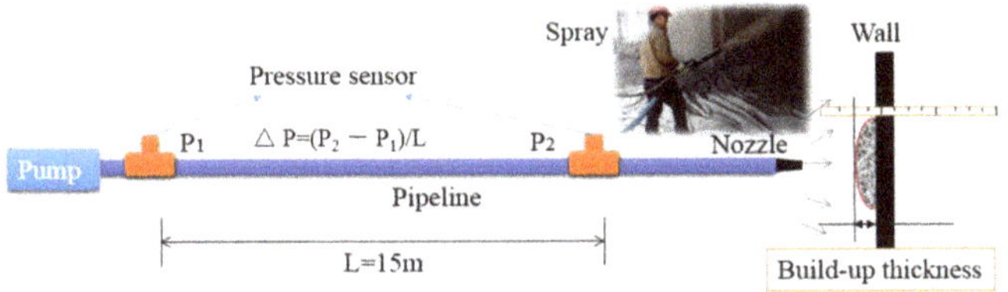

Figure 2. Pumping and shotcreting tests of fresh concrete.

3. Results and Discussion

The experimental results of pressure drop and build-up thickness are shown in Table 3.

Table 3. Results of pumpability and shootability tests of fresh concrete.

No.	Pressure Drop (MPa/m)	Build-Up Thickness (mm)	No.	Pressure Drop (MPa/m)	Build-up Thickness (mm)
WC45	0.09	170 ± 20	AE0.06%	0.03	130 ± 20
WC50	0.08	160 ± 20	WR0.3%	0.04	160 ± 20
WC55	0.04	140 ± 20	WR0.6%	0.04	150 ± 20
WC60	0.03	130 ± 20	WR0.9%	0.03	150 ± 20
SR50	0.07	$170 + 20$	WR1.2%	0.03	130 ± 20
SR60	0.05	$180 + 20$	SF5%	0.07	170 ± 20
SR70	0.09	160 ± 20	SF10%	0.08	170 ± 20
FI6	0.08	190 ± 20	SF15%	0.10	190 ± 20
FI12	0.07	190 ± 20	FA5%	0.05	180 ± 20
FI29	0.06	190 ± 20	FA10%	0.04	170 ± 20
AE0.02%	0.05	150 ± 20	FA15%	0.03	160 ± 20
AE0.04%	0.04	130 ± 20			

3.1. Influence of Various Factors on Rheology and Air Content of Fresh Wet Shotcrete

3.1.1. Effect of Water to Binder Ratio (WC)

Figure 3 describes the effect of WC on the properties of fresh wet shotcrete. When WC increases from 0.45 to 0.60, the slump and air content increase, while the rheological parameters (relative yield stress and viscosity) decrease. It can be explained that the larger the WC, the more the free water content in the fresh wet shotcrete, the thicker the outer paste coating of the particles, the larger the spacing between particles, and the bigger the fluidity. In addition, fresh concrete with higher fluidity creates more opportunities for the introduction of air bubbles into the concrete mixture [49]. On the contrary, these bubbles act as "ball effect" to improve the fluidity of fresh wet shotcrete. This schematic diagram is shown in Figure 4. These trends are similar to Yun's report [35], where they measured rheological properties using an IBB rheometer with an H-shaped impeller. The internal structure of the two rheometers is different, such as the shape of the rotor, rotation speed, the size of the rheometer and the distance between the inner and outer rotors. Hence, the relative rheological values measured in the two studies are different. However, the rheological parameters obtained in this study also provide meaningful practical knowledge for guiding the mix proportion of wet shotcrete.

It can be seen from Table 3 that with the increase of WC, the pressure drop decreases, i.e., smaller rheological parameters (yield stress and plastic viscosity) help to improve the pumping performance of fresh concrete. However, the high WC reduces the build-up thickness of concrete, and the shootability of fresh concrete with good pumping performance is poor. The author [35] also pointed out that high WC might be beneficial to pumping performance, but low viscosity and low yield stress might reduce the thickness of the build-up layer. From another point of view, Pfeuffer et al. [50] pointed out that based on the assumption of better aggregate embedding, the viscosity and yield stress of shotcrete need to be lower to minimize the rebound rate. As shown in Figure 5, the wet shotcrete with low fluidity has a strong cohesive force, which increases the accumulation thickness h_1. At the same time, we redefine F_1 as the force that prevents particles from entering the shotcrete matrix, because the F_1 is greater than the incident force F_2, it may cause certain rebound. On the contrary, wet shotcrete with high fluidity provides convenience for aggregate embedding and reduces rebound. However, it is easy to collapse, resulting in a very thin accumulation of h_2.

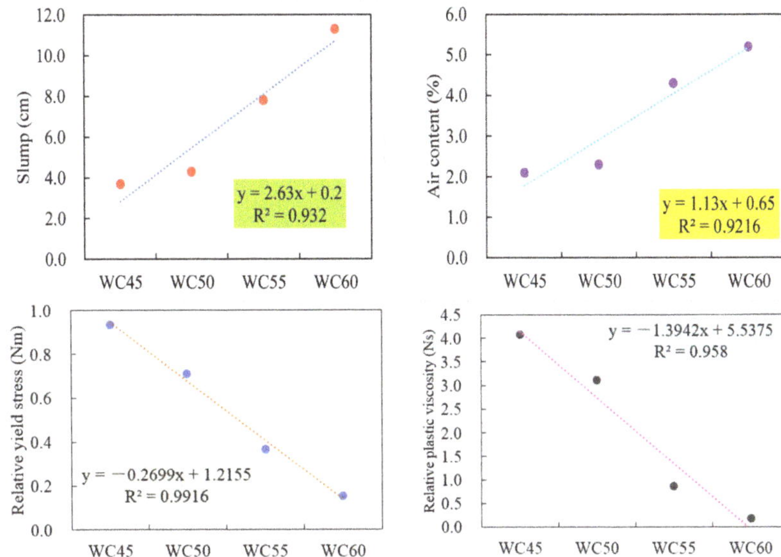

Figure 3. Effect of water–binder ratio on slump, air content, relative yield stress, and viscosity.

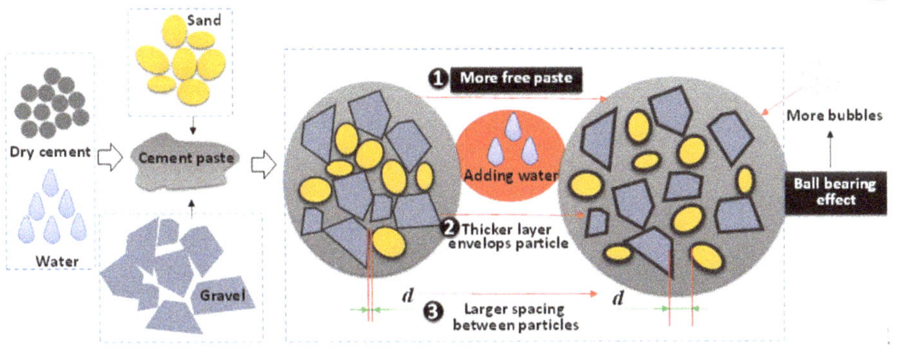

Figure 4. Schematic of the flowability change after increasing water content.

According to the fitting results of the linear relationship in Figure 3, the fitting effect of the four relationships between WC and rheological parameters is good, R^2 is more than 0.9, close to 1. The linear fitting effect between WC and yield stress is the best, R^2 = 0.9916. The WC vs. air content is the worst, R^2 = 0.9216. The bubble system in fresh wet shotcrete becomes unstable due to high WC, which may be the reason for poor fitting effect of air content.

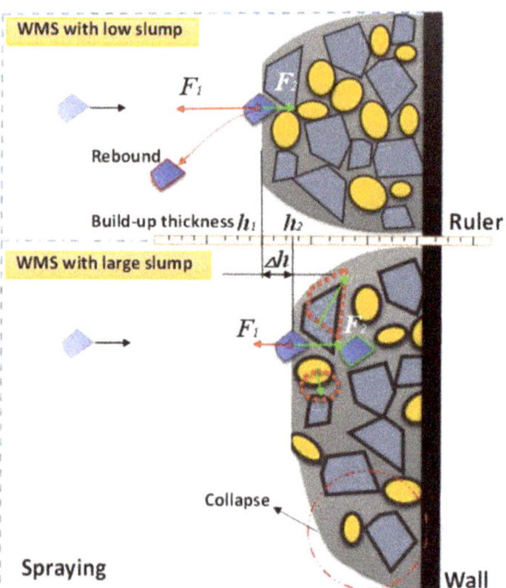

Figure 5. Shootability compression between low slump and large slump.

3.1.2. Effect of Sand Ratio

Figure 6 shows that with the increase of SR, the slump decreases gradually, while the air content increases. Within the range of 50–60% of SR, the relative yield stress increases, and the relative viscosity changes slightly with a downward trend. Some researchers [51–53] have reported the same result that increasing strontium increases the yield stress but decreases the plastic viscosity. Westerholm et al. [54] also found that natural fine aggregate has little effect on plastic viscosity.

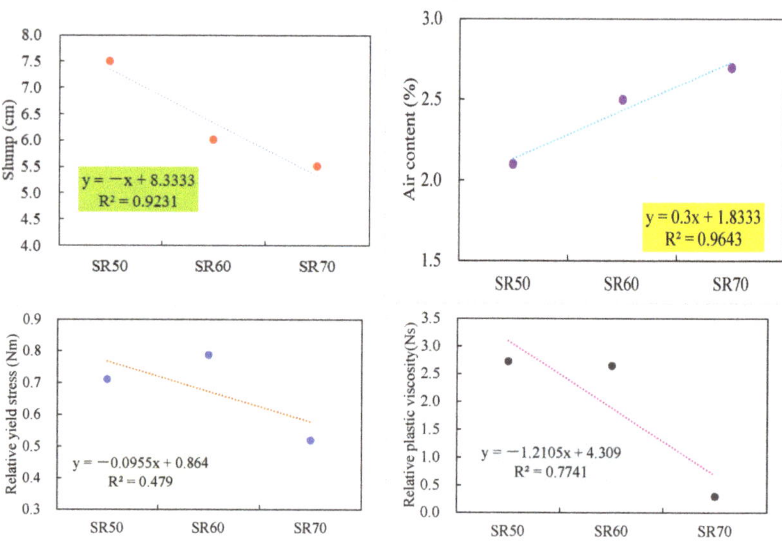

Figure 6. Effect of sand ratio on slump, air content, relative yield stress and viscosity.

The analysis shows that the texture of sand is smoother than that of gravel when "ball effect" is considered. The increase of SR reduces the content of gravel, and more sand as a kind of ball can fill the gap between coarse aggregates, resulting in the release of cement paste embedded in the voids and the decrease of inter particle friction. However, if the sand volume fraction is further increased, the total surface area of solid particles will increase, the consumption of cement paste will increase, the yield stress and friction between particles will increase, which will reduce the fluidity of fresh wet shotcrete. The author [55] divides the influence of SR on the fluidity of concrete into two parts: the smaller volume of SR has a negative effect on the fluidity of concrete, because there is no mortar filling the gap in the coarse aggregates; the higher volume SR also reduces the fluidity, because the high specific surface area of fine aggregate consumes too much free paste that lubricates the aggregate particles and the inner wall of the pipe. Therefore, there is a critical SR value, and the positive and negative effects of free paste or ball bearing are balanced. Considering the shootability, the round sand particles are unfavorable to the accumulation thickness. The greater the sand content, the greater the density of shotcrete and the higher the compressive strength.

According to Table 3, SR60 is the best value to obtain both good pumping performance and good shooting performance in the range of SR50 to SR70. Of course, due to the small number of experimental samples in this study, further research is needed to find the accurate critical point. The positive and negative effects of SR on liquidity of wet shotcrete are shown in Figure 7.

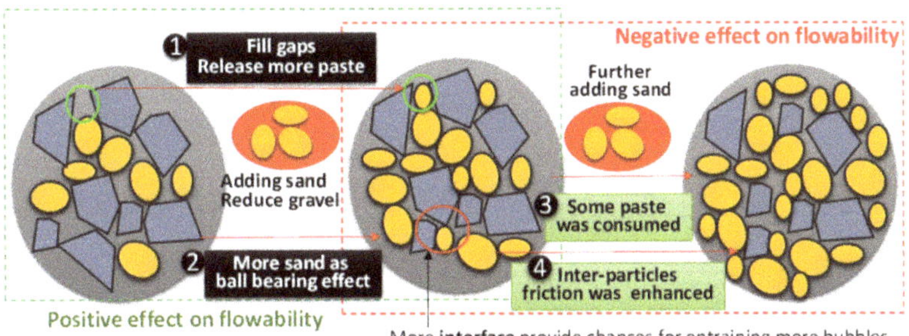

Figure 7. Schematic of the effect of sand ratio (SR) on the flowability.

In terms of air content, adding more fine aggregate will increase the total surface area of solid particles in fresh wet shotcrete, which provides more opportunities for liquid-solid interaction, thus bringing more bubbles [56,57]. It was noted that the mixture labeled "SR70" was mixed with 0.3% water-reducing agent to ensure good pumping effect. Therefore, when the content of SR is 60–70%, the yield stress and viscosity are reduced by further increasing SR. In addition, the change of SR affects the mix proportion and bulk density. Jiao et al. [58] pointed out that the volume fraction and bulk density of aggregate have obvious influence on the rheological properties of concrete. The bulk density is related to the type, gradation, and size distribution of aggregates. Relevant studies on these aggregate parameters are given in [51,58,59].

According to the linear fitting results in Figure 6, the fitting effect between SR and slump or air content is good, R^2 is greater than 0.9. However, the fitting of rheological parameters is poor, the relative yield stress $R^2 = 0.479$, the relative viscosity $R^2 = 0.7741$, which may be due to the introduction of superplasticizer in SR70.

3.1.3. Effect of Fiber

Figure 8 shows the relationship between fiber length and rheological properties, and air content. The air content, relative yield stress, and viscosity showed an increasing trend while slump decreases with the increase of fiber length. The results show that the fiber has an adverse effect on the fluidity with high yield stress and high viscosity, which is consistent with the results confirmed by Sivakumar et al. [60]. The reason may be that the specific surface area of long fiber is large, which can consume a part of free cement paste, and the friction between particles and aggregate increases with the increase of fiber length. Tabatabaian et al. [20] indicated the existence of fiber hinders the flow of cement paste. In the same way, increasing the total surface of the solid after mixing the long fibers will cause more bubbles to be entrained in, resulting in a higher air content in the "FI29 mm" mixing ratio. The schematic diagram of FI affecting fresh shotcrete characteristics is shown in Figure 9.

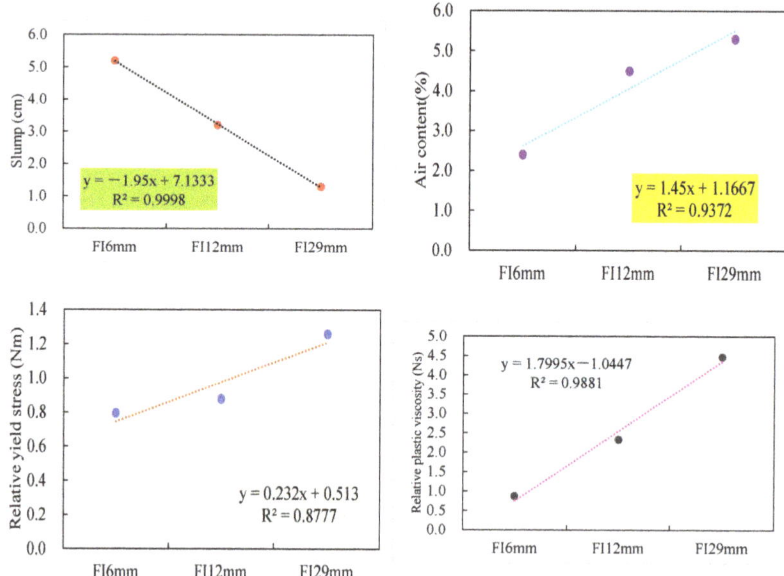

Figure 8. Effect of fiber on slump, air content, relative yield stress, and viscosity.

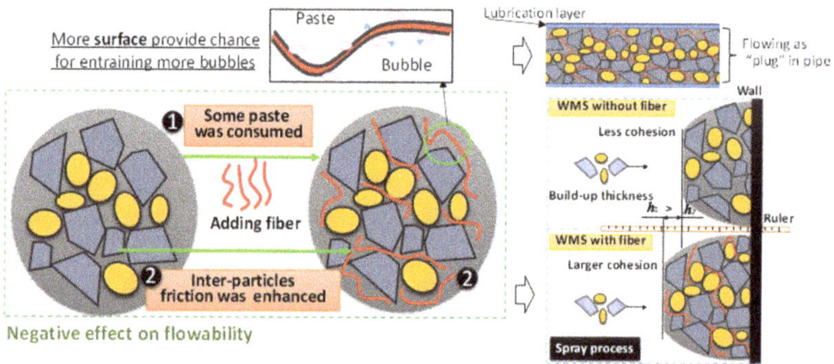

Figure 9. Schematic of fiber (FI) affecting the flowability, shootability, and air content.

Combined with the experimental results of pumpability and shootability in Table 3, longer fibers help to reduce pressure drop and increase build-up thickness. In addition, it is determined that the addition of fibers can improve the shootability by increasing the stacking thickness [35]. In report [32], the authors point out that low slump does not mean low pumpability. Adding fiber into fresh concrete can improve the bond strength of fresh concrete and reduce the segregation during pumping. However, Yun et al. [35] considered that the mixed fiber has an adverse effect on the pumpability of fresh wet shotcrete (reducing slump and increasing yield stress). Different results may be related to the lubrication layer in the pipeline. If the lubricating layer has been formed, the low slump concrete will flow in the form of "plug", and the friction resistance is small.

According to the linear fitting results in Figure 8, the fitting effect of fiber and slump is the best, $R^2 = 0.9998$, and the fitting effect of yield stress is the worst, $R^2 = 0.8777$. The prediction results of the influence of fiber on rheological properties and air content may be used in practice.

3.1.4. Effect of Air-Entraining Agent

Figure 10 shows that with the increase of air-entraining agent content, the air content slump increase, while the relative yield stress and viscosity decrease. The results show that adding more air-entraining agent in fresh wet shotcrete will produce a lot of bubbles, which plays a role of "ball effect". These entrained bubbles play a key role in lubricating aggregate and further affect the consistency of fresh wet shotcrete. The fluidity of wet shotcrete is improved, the slump is increased, and the yield stress and viscosity are reduced. Similarly, Yun et al. [35] has been pointed out that the torque viscosity decreases significantly with the increase of air content. However, in an earlier study [61], the author showed that there was a critical air content of 5%, before which both flow resistance and torque viscosity decreased; after this point, the two rheological parameters tended to be stable. In our study, when 0.02% air-entraining agent is mixed, the air content reaches 13.5%, far exceeding the critical value of 5%. The corresponding criterion in reference [61] is invalid. This difference can be explained by two factors: one is the application of different types of test equipment or the type of fresh concrete (ordinary concrete with aggregate size greater than 30 mm and shotcrete with maximum particle size of 10 mm); the other can be found in [58]. The author [58] proposed that since bubbles are attracted to the cement particles to form a bubble bridge, which increases the bond strength between cement particles (similar to flocculated particles), the entrained bubbles may lead to an increase in yield stress. Once the bubble bridge breaks under strong shear, some of the retained paste is released, apparently acting as a lubricant, as shown in Figure 11. We can conclude that bubbles have two functions: binding and lubrication. Both are related to shear strength and aggregate parameters. Therefore, some reports [62,63] show that the yield stress increases or remains stable with increasing air content.

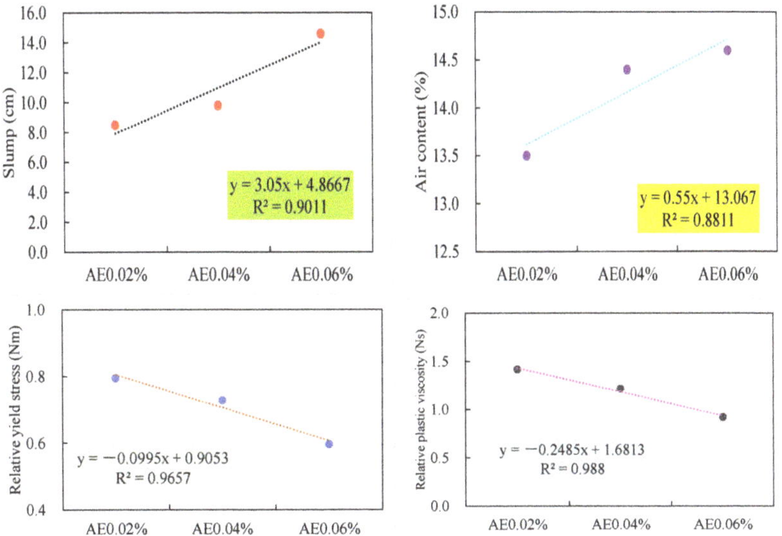

Figure 10. Effect of air-entraining agent on slump, air content, relative yield stress, and viscosity.

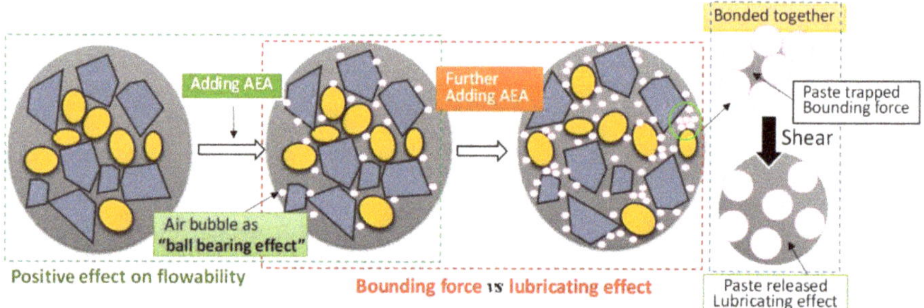

Figure 11. Schematic of the effect of air bubble on flowability of WET SHOTCRETE.

In our study, adding air-entraining agent can improve the pumping performance of concrete in Table 3. Therefore, we infer that in this measurement process, the lubrication effect of bubbles is significantly greater than that of adhesion, with the range of AE0.02%–AE0.06%. Due to the spraying process of fresh wet shotcrete, bubbles will break and disappear, this kind of fresh concrete has better shotcreting property with good cohesion. Therefore, the air-entraining agent can improve the pumpability and sprayability in a certain range, which is worthy of recommendation in the application of wet concrete.

According to the linear fitting results in Figure 10, the air-entraining agent has a good fitting effect on the rheological parameters, with the relative yield stress $R^2 = 0.9657$ and the relative viscosity $R^2 = 0.988$. Air-entraining agent has the worst fitting effect on air content, which may be the cause of instability of fresh concrete with more bubbles. The prediction results of total rheological properties and air content are ideal.

3.1.5. Effect of Water Reducer

Figure 12 shows that the relative yield stress, viscosity, and air content decrease with the increase of water reducer dosage from 0.3% to 1.2%, while the slump is opposite. Under the same water–binder ratio, the fluidity of fresh concrete is greatly improved when the quality of water-reducing agent is mixed. When 1.2% water-reducing agent is added, the slump value reaches 15.1 cm, which is much higher than that of wet shotcrete with other admixtures. There was no bleeding or separation during tests. Jiao et al. [58] indicated that the spatial resistance effect of water-reducing agent is one of the reasons for this phenomenon. After adding water reducer, the yield stress and viscosity of fresh concrete decrease significantly. As shown in Figure 13, after adding water reducer, the flocculability of cement particles decreases because water reducer makes cement particles bringing same charge. In addition, the efficiency of water reducer also depends on the adsorption of water-reducing agent molecules to cement particles and the repulsion force produced by adsorption molecules. Kwan et al. [64] found that the addition of water-reducing agent increases the effective layer thickness of cement particles covered by paste. The increase of the bridging distance between particles reduces the maximum attraction between particles. Perrot et al. [65] concluded that with the increase of water-reducing agent dosage, the average surface spacing increases, and the colloidal interaction and flocculation degree between particles are reduced, thus reducing the rheological parameters of fresh wet shotcrete.

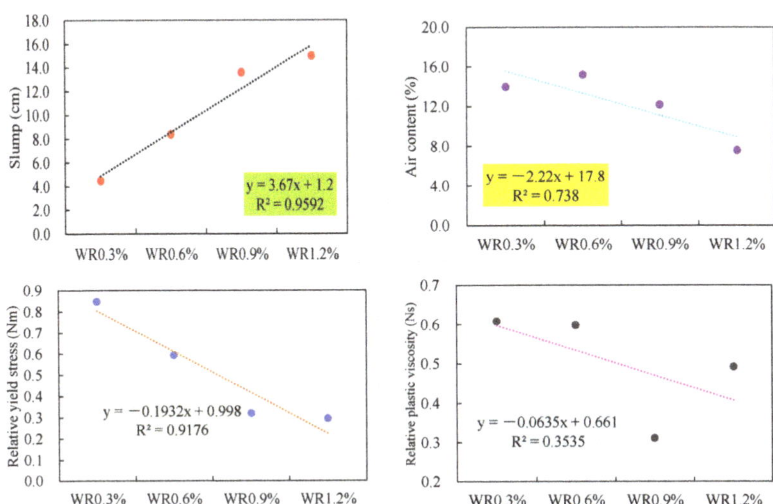

Figure 12. Effect of water reducer on slump, air content, relative yield stress, and viscosity.

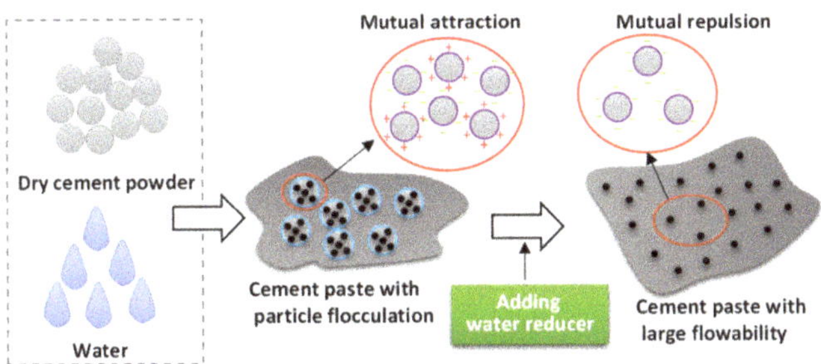

Figure 13. Mechanism of water reducer affecting cement paste.

When WR exceeds 0.9%, the relative yield stress decreases slowly, the plastic viscosity increases inversely. It might be explained that there is a critical saturation point for the efficiency of superplasticizer. Beyond this point, more superplasticizers will increase the viscosity of the liquid. Compared with other admixtures (except air-entraining agent), the air content of wet shotcrete with water-reducing agent is relatively high, with an average of 12.25%. The analysis shows that the water reducer contains a certain amount of surfactant and thus introduces some bubbles. If the superplasticizer is continued to be added, excessive water reducer will increase the viscosity of the liquid and hinder the generation of bubbles, which will lead to the decrease of air content.

According to the fitting results in Figure 12, the slope of the linear equation represents the rate of increase or decrease. The decrease of relative yield stress is greater than that of relative plastic viscosity. The results show that superplasticizer has good fitting effect on slump and relative yield stress, $R^2 = 0.9592$ and $R^2 = 0.9176$, respectively, which can be used to predict the changes of these two properties.

Table 3 shows that adding more superplasticizers can reduce the stacking thickness. It is pointed out that although water-reducing agent is almost used in modern pouring concrete due to its economic and technical benefits, the application of water-reducing agent is not conducive to improving the shootability of concrete. It is suggested that when premixing water-reducing agent in front of pump, additives such as viscosity agent or accelerator should be added at the same time to restore or improve its shooting performance.

3.1.6. Effect of Fly Ash

As shown in Figure 14, with the increase of fly ash content from 5% to 15%, the slump value increases more than twice, and the relative yield stress and viscosity decrease by about 50%. Generally speaking, the fluidity of shotcrete can be significantly improved by adding fly ash. Similar results with positive effect of fly ash on fluidity are also reported in references [66,67]. According to the pumping test in Table 3, adding fly ash can improve the pumping performance, reduce the pressure drop, but reduce the sprayability, and the build-up thickness is small. This phenomenon may also be attributed to the "ball effect" and "free paste effect". The content of CaO in the fly ash used in this test is less than 10%, which belongs to Class F fly ash. The fly ash particles have spherical geometry and smooth surface (Figure 15 [68]) to enhance the "ball bearing effect" by reducing friction and promoting sliding between angular particles. Vance et al. [69] pointed out that this effect will be amplified with the widening of fly ash particle size distribution. Fly ash, as a fine particle with a particle size of 0.4–100 μm, fills the gap between larger particles, and extrudes the original water in the gap to increase the volume of free paste, which is called free paste effect. The use of fly ash instead of cement reduces the cement concentration and further reduces the flocculation of cement particles. In addition, because the specific gravity of fly ash particles is lower than that of cement particles, the volume of paste will increase when fly ash replaces cement with the same quality. Therefore, more free paste volume as lubricant increases the fluidity of wet shotcrete, which is called dilution effect. The influence of fly ash on the fluidity of fresh concrete is shown in Figure 15.

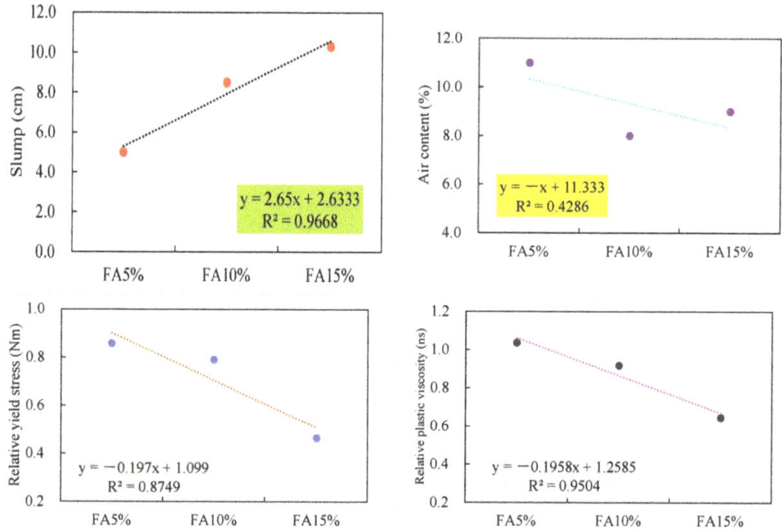

Figure 14. Effect of fly ash on slump, air content, relative yield stress, and viscosity.

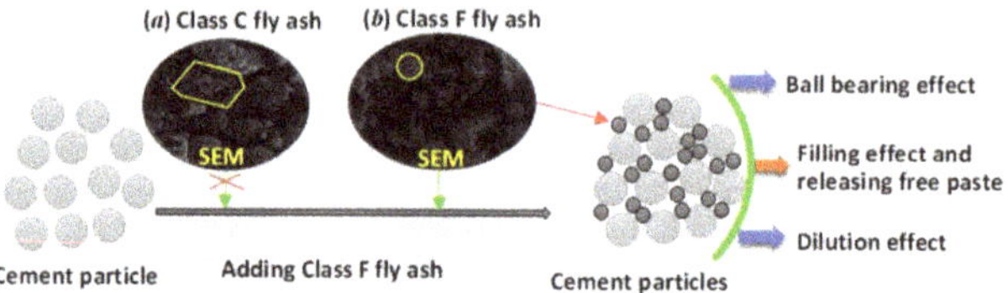

Figure 15. Effect of fly ash on the flowability of fresh wet shotcrete.

This does not mean that the more fly ash is, the higher the fluidity of the new concrete. Laskar et al. [66] found that when a higher content of fly ash is mixed, the yield stress increases. This is also related to the types of fly ash, for example, the Class C fly ash in Figure 15a is irregular and porous, and the lubrication is weak. CaO in Class C fly ash is more, and its viscosity is relatively higher after hydration.

According to the fitting results in Figure 14, the fitting effect of fly ash on slump and relative plastic viscosity is good, $R^2 = 0.9668$, $R^2 = 0.9504$ respectively, with good prediction effect. With the increase of fly ash content, the air content changes greatly, leading to the bad fitting effect of fly ash and air content, R^2 is less than 0.5.

3.1.7. Effect of Silica Fume

Figure 16 shows that the addition of silica fume reduces the fluidity of fresh wet shotcrete, reduces the slump, and increases the relative yield stress and viscosity. In terms of filling effect and ball bearing effect, the average particle size of silica fume (about 0.1–0.3 μm) is smaller than that of cement (30–60 μm). Jiao et al. [58] pointed out that the particles below 0.1 μm account for more than 80% in silica fume, and the particle geometry is circular. These conditions provide the possibility of filling silicon powder particles between larger particle gaps as ball bearings. In this normal case, the free paste effect enhanced by particle filling effect will improve the fluidity of fresh wet shotcrete. This mechanism is similar to that of fly ash. However, due to the high silica content (85–96%), silica fume has high chemical activity. It participates in the hydration process of cement and increases the consistency of cement and the friction between particles. After adding silica fume, the fluidity of fresh wet shotcrete decreases because the effect of free paste is less than the effect of flocculation. In this study, the relative yield stress and viscosity increased after adding 5% silica fume. This result is similar to previous work [35], which reported that the flow resistance of fresh concrete increases with silica fume content greater than 7%.

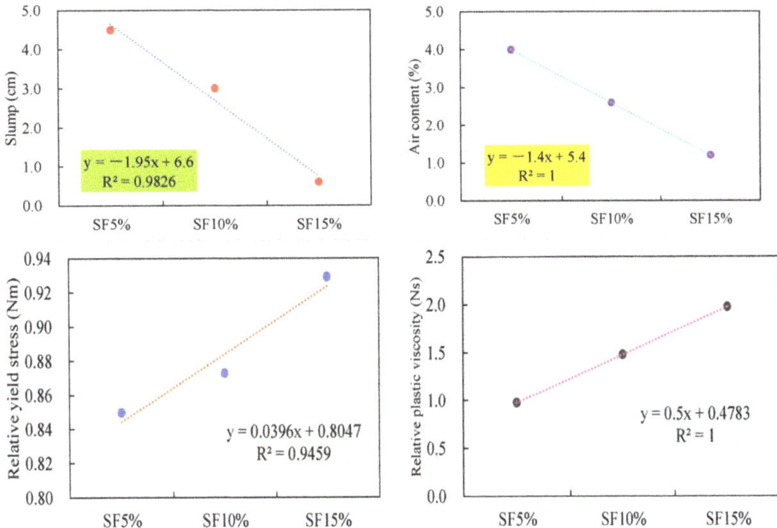

Figure 16. Effect of silica fume on slump, air content, relative yield stress, and viscosity.

Considering that silica fume reduces the fluidity of wet shotcrete, superplasticizer was added into the mixture to ensure good pumping performance. It can be seen from Figure 16 that the effect of water-reducing agent on wet shotcrete with silica fume is not obvious, because after adding water reducer, the slump cannot reach a large slump, and the slump value is less than 5 cm. It has been reported in the previous literature [35,38,66] that the high specific surface area of silica fume adsorbs water-reducing agent molecules and reduces the water-reducing effect. Air content decreases with the increase of silica fume content. This may be caused by the decrease of water because the high fineness and high chemical activity of silica fume increase the demand for water. Under the same water content, adding more silica fume consumes too much water and reduces the chance of introducing bubbles into fresh shotcrete.

As shown in Table 3, the increase of yield stress and viscosity is beneficial to build-up thickness. Due to the additional flocculation of silica fume, the bond strength of concrete mixture on surface sprayed is enhanced, and the segregation and bleeding in the pumping process are reduced. Beauprep [31] gave a similar result, i.e., when 10–15% silica fume is added to wet shotcrete, silica fume has a positive effect on both sprayability and pumpability. It can be seen from the fitting results in Figure 16 that the silica fume has a good fitting effect on the air content and rheological parameters.

We try to illustrate the influence of various factors on shotcrete performance through pictures, such as Figures 4, 5 and 7, etc. It should be noted that only three or four factor levels are selected in this study. Figures 3, 6, 8, 10, 12, 14 and 16 only adopt linear fitting method, but in fact, exponential fitting equation may be more suitable for the change trend of some influencing factors, such as w/c ration. Because the test data of each factor is relatively small and the fitting variance is small, all the fitting patterns in those figures are only used as a reference. In future studies, we will increase the number of test samples.

3.2. Difference Analysis of Various Impact Factors

According to the experimental data of rheological parameters and air content, the influence degree of rheological properties and air content are analyzed within the range of variables set in this test. To eliminate the influence of different factors on the measurement scale, the "coefficient of variation" analysis method is adopted [70,71]. Calculation steps: (1) calculate the average value and standard deviation of the same project affected by a different degree of influencing factors; (2) divide the standard deviation by the average value, and the result is "coefficient of variation"; (3) plot the coefficient of variation into a graph. The corresponding results are shown in Figures 17–20. The greater the coefficient of variation, the greater the influence of each factor on wet shotcrete.

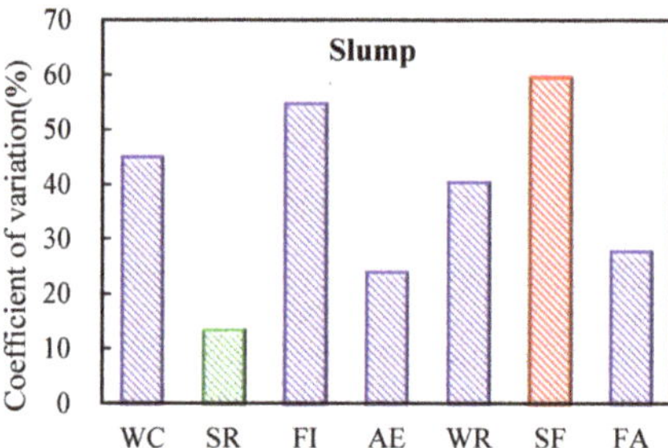

Figure 17. Difference analysis of factors on slump.

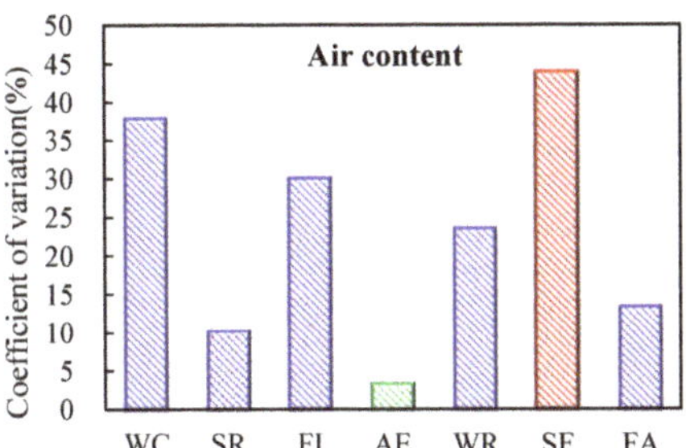

Figure 18. Difference analysis of factors on air content.

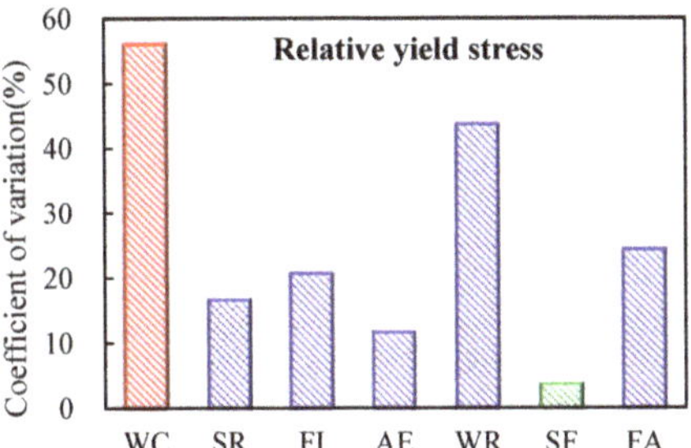

Figure 19. Difference analysis of factors on yield stress.

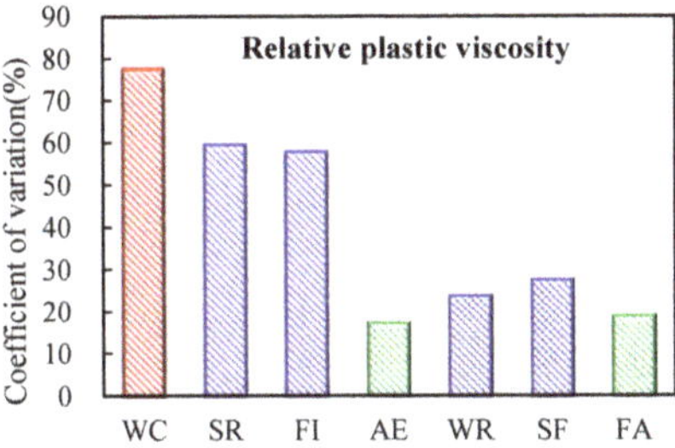

Figure 20. Difference analysis on viscosity.

It can be seen from Figures 17 and 18 that silica fume has the greatest influence on the slump and air content within a given range of variables, with variation coefficients of 60.1% and 45.2% respectively. Sand ratio has the least effect on slump (13.2%) while air-entraining agent has the least effect on gas content (4.3%). In addition, water–binder ratio (WC), fiber (FI) and water-reducing agent (WR) have great influence on slump and air content.

It can be seen from Figures 19 and 20 that the water–binder ratio (WC) has the greatest effect on the relative yield stress and plastic viscosity within a given range of variables. Secondly, for yield stress, water reducer (WR) and fly ash (FA) have great influence on it, silica fume (SF) has the least influence; for plastic viscosity, sand ratio (SR) and fiber (FI) have great influence on it, while air-entraining agent (AE) or fly ash (FA) has relatively small effect.

In fact, the coefficient of variation of influencing factors is related to the range of variables set in this experiment, and related to the special working performance of the same influencing factor. For example, fly ash has "ball effect" and silica fume has "filling effect". It is hoped that the analysis of the coefficient of variation can provide a useful reference for the mix proportion design of wet shotcrete.

3.3. Comparison of the Effects of Fly Ash, Silica Fume on Rheological Properties and Air Content

To understand deeply the effects of silica fume or fly ash on the properties of fresh wet shotcrete, the comparison between these two and other admixtures were analyzed. According to the comparative analysis of Figure 21, it can be concluded that:

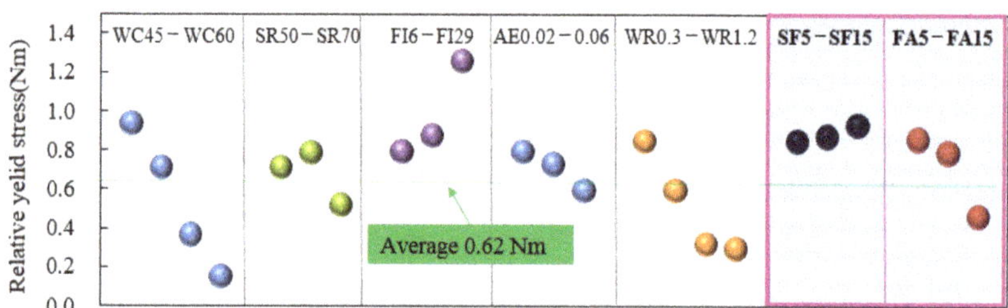

(**a**) Effect on relative yield stress

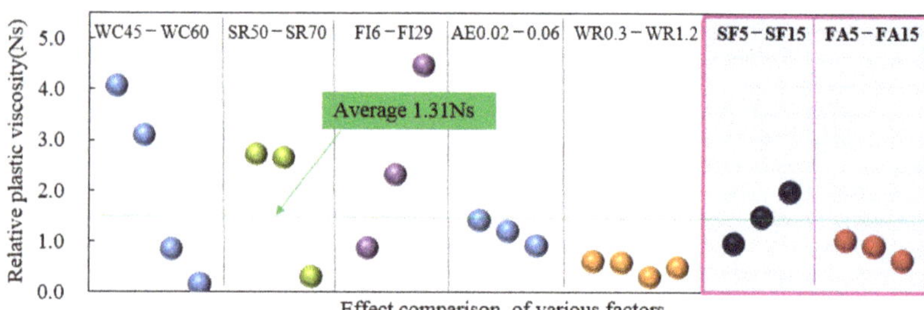

(**b**) Effect on relative plastic viscosity

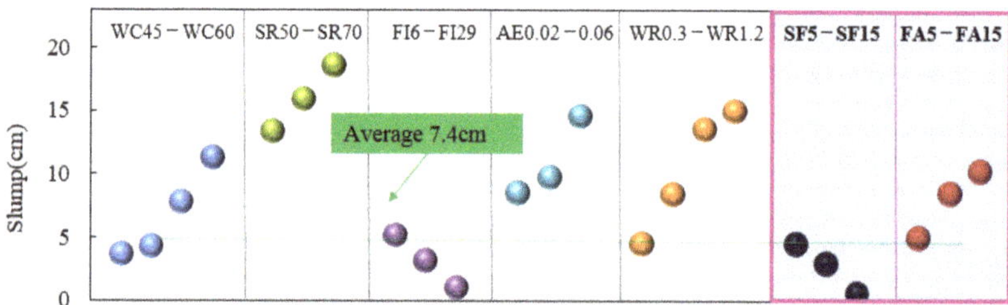

(**c**) Effect on slump

Figure 21. *Cont.*

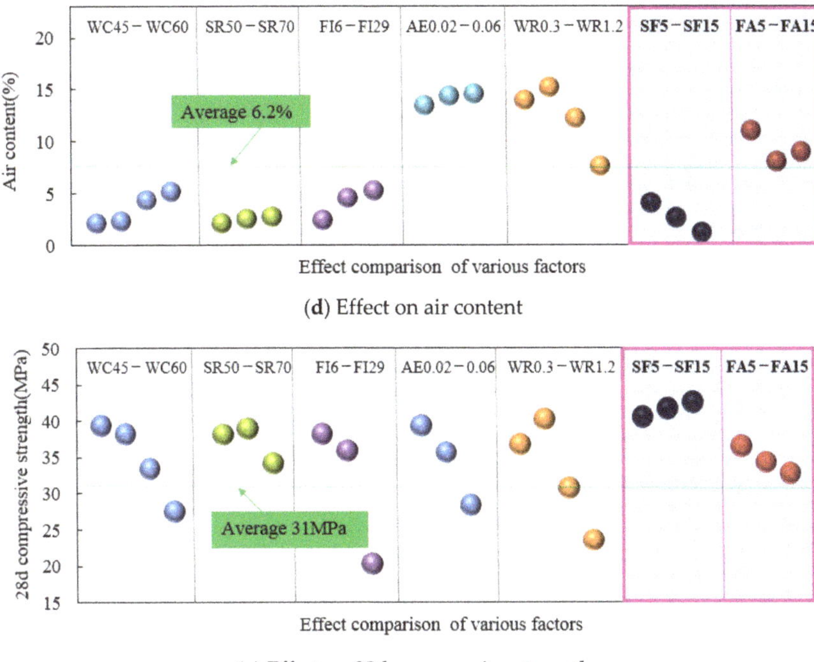

(**d**) Effect on air content

(**e**) Effect on 28d compressive strength

Figure 21. Comparative analysis of influence between fly ash, silica fume, and other factors.

In the aspect of relative yield stress, the effect of silica fume (SF) on the yield stress of wet shotcrete is similar to that of low water–binder ratio (WC), medium sand ratio (SR) and fiber (FI), almost exceeding the average value of 0.745 Nm. Fly ash (FA) and air-entraining agent (AE) have the same effect, showing a downward trend. It can be seen from the data in Table 3 that the high yield stress is beneficial to the shoot property, and the addition of silica fume or a small amount of fly ash can produce a better build-up thickness.

In the aspect of relative plastic viscosity, the influence degree of fly ash on plastic viscosity is similar to that of air-entraining agent, which is lower than the average value of 1.31 Ns. The similar effect can be produced by low dosage of water-reducing agent, high water–binder ratio, or high sand ratio. When the dosage of silica fume is 5–15%, the plastic viscosity is little affected and fluctuates around the average value. For the flow of fresh wet shotcrete in the pipeline, if a lubricating layer is formed on the inner wall of the pipe, the higher viscosity will improve the anti-segregation and anti-bleeding ability of the wet shotcrete containing silica fume, avoid blocking and increase the build-up thickness. However, for shear flow, lower viscosity is beneficial to the pumping performance of fresh wet shotcrete. Therefore, adding fly ash is beneficial to pumping.

In terms of slump: the effect of adding silica fume and fiber in fresh wet shotcrete is similar, and the slump of both is lower than the average value of 7.4 cm. The effect of adding fly ash on slump is similar to that of low air-entraining agent or low water-reducing agent. According to the test of pumpability and shootability, high slump is beneficial to pumping. Fly ash is better than silica fume in pumping process, but the shootability is opposite. It should be noted that the change trend of slump is opposite to that of relative yield stress, which is also reported in [29,52].

In terms of air content, the air content of wet shotcrete mixed with silica fume is lower than the average value of 6.2%, which is similar to that of fiber, low water–binder ratio, or high sand ratio. However, fly ash can produce similar effect of air-entraining agent or water-reducing agent at high air content. If the assumption that spraying behavior will reduce air content is true, fresh wet shotcrete with high air content is beneficial to pumping and spraying. Therefore, it seems that adding silica fume is not conducive to the air content of wet shotcrete, while fly ash with relatively high air content is better.

Compressive strength: the influence of fly ash content on compressive strength is moderate, slightly higher than the average value. Considering the pumping and shotcreting performance of fly ash, fly ash is undoubtedly the common admixture of wet shotcrete. From the rheological parameters and air content, although the fluidity of concrete mixed with silica fume is low, it has great advantages in strength and sprayability. In some special projects that need to add silica fume into shotcrete, the pumpability can be improved by adding water-reducing agent or air-entraining agent.

3.4. Summary of Influence of Admixtures or Additives on Pumpability and Shootability

According to the above experimental analysis, free paste and round particles are the key factors affecting the pumpability and sprayability of wet shotcrete. As shown in Figure 22, cement paste has three functions: coating particles, filling gaps, and lubricating pipe wall. The "ball effect" and "free paste effect" have compound effects on the properties of wet shotcrete. In theory, these two effects, i.e., enough free paste or enough balls, can reduce the yield stress and improve the fluidity of concrete, which is called positive effect. However, in the fresh concrete system, if the free cement paste or smooth ball is insufficient, it is called negative effect, which will reduce the fluidity of fresh concrete. For example, in Figure 7, it can be concluded (from SR50 to SR60) that the negative paste effect is greater than the positive ball effect, resulting in an increase in yield stress and a decrease in slump.

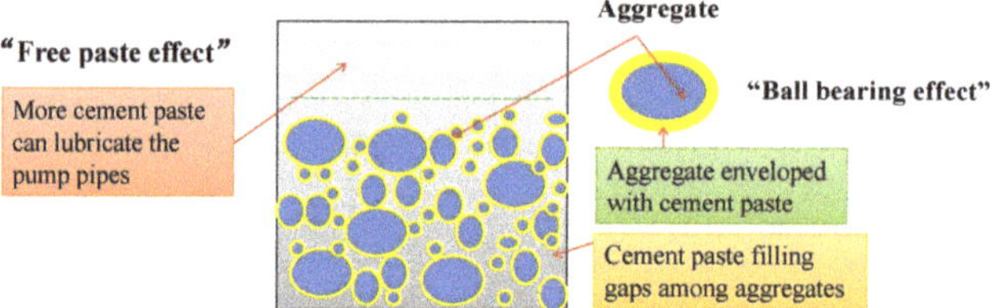

Figure 22. Free paste effect and ball bearing effect in fresh concrete system.

Yammine et al. [72] also found that the rheological properties of fresh concrete changed significantly between hydrodynamic and friction control. When the cement paste volume is small, the fluidity is mainly determined by the direct friction between particles. When the cement paste volume is large, the interaction between fluid and particles or between particles is hydrodynamic. The movement of the particles causes the paste to flow, where additional energy is dissipated. Therefore, with the increase of paste volume fraction, the plastic viscosity decreases. The additional paste volume significantly reduces the negative effect of coarse aggregate on the fluidity of fresh concrete [54]. Various factors affecting shotcrete properties are summarized in Table 4.

According to the above test results, to meet the different workability, such as pumping, spraying, or compressive strength, the general mix proportion range is given. Finally, the optimum mix proportion is recommended, as shown in Table 5.

Table 4. Summary of various factors affecting shotcrete in pumpability and shootability.

Variable	Flowability Factor		Effect on Wet Shotcrete	
	Ball Bearing Effect	Free Paste Effect or Filling Effect	Pumpability	Shootability
Increasing WC	Creating possibility for entraining bubbles as ball bearing effect	Producing more free paste	Positive at some certain level	Negative for build-up thickness; Positive for reducing rebound
Increasing sand ratio	Sand as ball bearing effect	Sand fills gaps to release more paste	Positive	Negative for build-up thickness; Positive for strength
Fiber	None	Adding fiber reduces free paste	Positive due to avoiding bleeding and segregation	Positive for build-up thickness
Air-entraining agent	Bubbles as ball bearing effect	When bridge is broken, paste was released	Positive	Positive when bubbles were broken during spraying
Water reducer	None	Releasing free cement paste	Positive	Negative
Fly ash	Spherical geometry and smooth surface as ball bearing effect	Dilution effect with lower specific gravity; filling effect with small particle size	Positive	Negative at some degree
Silica fume	Spherical geometry as ball bearing effect	Filling effect with small particle; flocculation reducing paste	Positive for reducing bleeding and segregation	Positive for build-up thickness

Table 5. Mixture proportions recommended for meeting various shotcrete requirements.

Requirements	WC (w/b)	Water (kg/m^3)	SR(%)	Sand (kg/m^3)	FI (mm)	AEA (%)	WR (%)	SF (%)	FA (%)
Pumpability	0.55–0.60	220	60	1000	12–29	0.04–0.06	0.9–1.2	15	5
Shootability	0.45–0.50	220	50–60	1000	12–29	0.02–0.04	0.3–0.6	5–10	15
Compressive strength	0.45	220	60	1000	6	0.02	0.6	15	5
Optimum mix proportion for wet shotcrete	0.5	220	60	1000	12	0.04	0.6	10	10

4. Conclusions

To investigate the influence of various admixtures on the fresh performance of shotcrete, the rheological properties and air content were measured. Through the comparative analysis of influencing factors and different analysis, it is helpful to understand the mix proportion design of fresh wet shotcrete.

(1) When WC increases, the slump and air content increase generally, while the rheological parameters decrease. with the increase of SR, the slump decreases gradually, while the air content and the relative yield stress increase. The air content, rheological parameters showed an increasing trend, while slump decreases with the increase of fiber length. With the increase of air-entraining agent content, the air content and slump increase, while rheological parameters decrease. Rheological parameters and air content decrease with the increase of superplasticizer dosage while the slump is opposite. With the increase of fly ash, the slump value increases, and rheological parameters decrease. The addition of silica fume reduces the slump, and increases the rheological parameters.

(2) With the increase of WC, the pressure drop decreases, i.e., smaller rheological parameters help to improve the pumping performance of fresh concrete. However, the high WC reduces the shootability. SR60 is the best value to obtain both good pumping performance and good shooting performance. Longer fibers help to reduce pressure drop and increase build-up thickness. The air-entraining agent can improve the pumpability

and sprayability in a certain range. The increase of yield stress and viscosity is beneficial to build-up thickness. It was found that here are two key factors affecting the fluidity and shotcreting of fresh concrete: free paste effect and ball effect. These two positive effects are conducive to the pumpability of fresh concrete.

(3) Through comparative analysis, the effect of silica fume on yield stress is similar to that of fiber and medium sand ratio, and the influence degree of silica fume on plastic viscosity fluctuates around the average level. The effect of silica fume on air content is similar to that of fiber or high sand ratio. The effect of fly ash is basically the same as that of air-entraining agent on rheological properties or air content. The effect of silica fume and fly ash on slump is opposite to that of relative yield stress.

(4) Within a given range of variables, silica fume has the greatest influence on the slump and air content of fresh wet shotcrete. The water–binder ratio has the greatest influence on the relative yield stress and plastic viscosity. Finally, we concluded the summary of various factors affecting shotcrete in pumpability and shootability. Mixture proportions recommended for meeting various shotcrete requirements were advised. To study the rheological properties of shotcrete more accurately, we will add more experimental variables in the next step.

Author Contributions: Conceptualization, J.X. and G.L.; methodology, X.C.; formal analysis, N.G.; investigation, N.G. and G.L.; resources, G.L.; data curation, X.C.; writing—original draft preparation, G.L.; writing—review and editing, J.X. and X.C.; funding acquisition, G.L. All authors have read and agreed to the published version of the manuscript

Funding: This research was funded by National Key Research and Development Plan of China, grant number 2017YFC0805203, National Natural Science Foundation of China, grant number 51974177, 51934004, 51604163, Natural Science Foundation of Shandong, grant number ZR201801280006, ZR2019QEE007, ZR2019MEE115, Exchange Projects of the 43rd Routine Session of the China-Czech Committee for Scientific and Technological Cooperation, grant number 43-4 and Special funds for Taishan scholar project.

Institutional Review Board Statement: Not applicable.

Informed Consent Statement: Not applicable.

Data Availability Statement: Not applicable.

Conflicts of Interest: The authors declare that they have no known competing financial interests or personal relationships that could have appeared to influence the work reported in this paper.

References

1. Choi, P.; Yeon, J.H.; Yun, K.-K. Air-void structure, strength, and permeability of wet-mix shotcrete before and after shotcreting operation: The influences of silica fume and air-entraining agent. *Cem. Concr. Compos.* **2016**, *70*, 69–77. [CrossRef]
2. Chen, L.; Li, P.; Liu, G.; Cheng, W.; Liu, Z. Development of cement dust suppression technology during shotcrete in mine of China-A review. *J. Loss Prev. Process. Ind.* **2018**, *55*, 232–242. [CrossRef]
3. Liu, G.; Cheng, W.; Chen, L. Investigating and optimizing the mix proportion of pumping wet-mix shotcrete with polypropylene fiber. *Constr. Build. Mater.* **2017**, *150*, 14–23. [CrossRef]
4. Hu, Z.-X.; Hu, X.-M.; Cheng, W.-M.; Zhao, Y.-Y.; Wu, M.-Y. Performance optimization of one-component polyurethane healing agent for self-healing concrete. *Constr. Build. Mater.* **2018**, *179*, 151–159. [CrossRef]
5. Chen, L.; Ma, G.; Liu, G.; Liu, Z. Effect of pumping and spraying processes on the rheological properties and air content of wet-mix shotcrete with various admixtures. *Constr. Build. Mater.* **2019**, *225*, 311–323. [CrossRef]
6. Chen, L.; Zhang, X.; Liu, G. Analysis of dynamic mechanical properties of sprayed fiber-reinforced concrete based on the energy conversion principle. *Constr. Build. Mater.* **2020**, *254*, 119167. [CrossRef]
7. Liu, G.; Cheng, W.; Chen, L.; Pan, G.; Liu, Z. Rheological properties of fresh concrete and its application on shotcrete. *Constr. Build. Mater.* **2020**, *243*, 118180. [CrossRef]
8. *Science and Technology of Concrete Admixtures*; Elsevier: Amsterdam, The Netherlands, 2015.
9. Dils, J.; Boel, V.; De Schutter, G. Influence of cement type and mixing pressure on air content, rheology and mechanical properties of UHPC. *Constr. Build. Mater.* **2013**, *41*, 455–463. [CrossRef]
10. Chen, J.J.; Kwan, A.K.H. Superfine cement for improving packing density, rheology and strength of cement paste. *Cem. Concr. Compos.* **2012**, *34*, 1–10. [CrossRef]

11. Koehler, E.; Fowler, D. Development of a Portable rheometer for fresh Portland cement concrete. In *Fiber Reinforced Concrete*; International Center for Aggregates Research: Austin, TX, USA, 2004; pp. 65–84.
12. Choi, M.S.; Kim, Y.J.; Jang, K.P.; Choi, M.S. Effect of the coarse aggregate size on pipe flow of pumped concrete. *Constr. Build. Mater.* **2014**, *66*, 723–730. [CrossRef]
13. Xu, W.; Zhang, Y.; Zuo, X.; Hong, M. Time-dependent rheological and mechanical properties of silica fume modified cemented tailings backfill in low temperature environment. *Cem. Concr. Compos.* **2020**, *114*, 103804. [CrossRef]
14. Park, C.K.; Noh, M.H.; Park, T.H. Rheological properties of cementitious materials containing mineral admixtures. *Cem. Concr. Res.* **2005**, *35*, 842–849. [CrossRef]
15. Rahman, M.K.; Baluch, M.H.; Malik, M.A. Thixotropic behavior of self compacting concrete with different mineral admixtures. *Constr. Build. Mater.* **2014**, *50*, 710–717.
16. Burroughs, J.F.; Weiss, J.; Haddock, J.E. Influence of high volumes of silica fume on the rheological behavior of oil well cement pastes. *Constr. Build. Mater.* **2019**, *203*, 401–407. [CrossRef]
17. Yun, K.K.; Choi, P.; Yeon J, H. Rheological characteristics of wet-mix shotcrete mixtures with crushed aggregates and mineral admixtures. *Ksce J. Civ. Eng.* **2017**, *5*, 1–11. [CrossRef]
18. Yun, K.-K.; Choi, P.; Yeon, J.H. Correlating rheological properties to the pumpability and shootability of wet-mix shotcrete mixtures. *Constr. Build. Mater.* **2015**, *98*, 884–891. [CrossRef]
19. Secrieru, E.; Fataei, S.; Schröfl, C.; Mechtcherine, V. Study on concrete pumpability combining different laboratory tools and linkage to rheology. *Constr. Build. Mater.* **2017**, *144*, 451–461. [CrossRef]
20. Tabatabaeian, M.; Khaloo, A.; Joshaghani, A.; Hajibandeh, E. Experimental investigation on effects of hybrid fibers on rheological, mechanical, and durability properties of high-strength SCC. *Constr. Build. Mater.* **2017**, *147*, 497–509. [CrossRef]
21. Burns, D. *Characterization of Wet shotcrete for Small Line Pumping*; Citeseer: Laval University, Quebec, QC, Canada, 2008.
22. Feys, D.; Khayat, K.H.; Khatib, R. How do concrete rheology, tribology, flow rate and pipe radius influence pumping pressure? *Cem. Concr. Compos.* **2016**, *66*, 38–46. [CrossRef]
23. Gołaszewski, J. Influence of mortar volume on rheological properties: Optimizing the workability of fresh high-performance concretes. *Betonw. Fert.-Tech.* **2008**, *74*, 44–51.
24. Feys, D.; De Schutter, G.; Verhoeven, R. Parameters influencing pressure during pumping of self-compacting concrete. *Mater. Struct.* **2012**, *46*, 533–555. [CrossRef]
25. Kwon, S.H.; Park, C.K.; Jeong, J.H.; Jo, S.D.; Lee, S.H. Prediction of Concrete Pumping: Part II—Analytical Prediction and Experimental Verification. *ACI Mater. J.* **2013**, *110*, 657–667.
26. Zhang, J.; Liu, L.; Cao, J.; Yan, X.; Zhang, F. Mechanism and application of concrete-filled steel tubular support in deep and high stress roadway. *Constr. Build. Mater.* **2018**, *186*, 233–246. [CrossRef]
27. Liu, G.; Guo, X.; Cheng, W.; Chen, L.; Cui, X. Investigating the migration law of aggregates during concrete flowing in pipe. *Constr. Build. Mater.* **2020**, *251*, 119065. [CrossRef]
28. Secrieru, E.; Cotardo, D.; Mechtcherine, V.; Lohaus, L.; Schröfl, C.; Begemann, C. Changes in concrete properties during pumping and formation of lubricating material under pressure. *Cem. Concr. Res.* **2018**, *108*, 129–139. [CrossRef]
29. Wallevik, O.H.; Wallevik, J.E. Rheology as a tool in concrete science: The use of rheographs and workability boxes. *Cem. Concr. Res.* **2011**, *41*, 1279–1288. [CrossRef]
30. Wang, J.; Niu, D.; Zhang, Y. Mechanical properties, permeability and durability of accelerated shotcrete. *Constr. Build. Mater.* **2015**, *95*, 312–328. [CrossRef]
31. Beaupre, D. Rheology of High Performance Shotcrete. Ph.D. Thesis, University of British Columbia, Vancouver, BC, Canada, 1994. [CrossRef]
32. Cheng, W.; Liu, G.; Chen, L. Pet Fiber Reinforced Wet-Mix Shotcrete with Walnut Shell as Replaced Aggregate. *Appl. Sci.* **2017**, *7*, 345. [CrossRef]
33. Li, P.; Zhou, Z.; Chen, L.; Liu, G.; Xiao, W. Research on Dust Suppression Technology of Shotcrete Based on New Spray Equipment and Process Optimization. *Adv. Civ. Eng.* **2019**, *2019*, 1–11. [CrossRef]
34. Chen, L.; Liu, G. Airflow-Dust Migration Law and Control Technology Under the Simultaneous Operations of Shotcreting and Drilling in Roadways. *Arab. J. Sci. Eng.* **2018**, *44*, 4961–4969. [CrossRef]
35. Yun, K.-K.; Choi, S.-Y.; Yeon, J.H. Effects of admixtures on the rheological properties of high-performance wet-mix shotcrete mixtures. *Constr. Build. Mater.* **2015**, *78*, 194–202. [CrossRef]
36. Pan, G.; Li, P.; Chen, L.; Liu, G. A study of the effect of rheological properties of fresh concrete on shotcrete-rebound based on different additive components. *Constr. Build. Mater.* **2019**, *224*, 1069–1080. [CrossRef]
37. ACI Committee. *ACI-506R Guide to Shotcrete, 506 ed.*; ACI Committee: Farmington Hills, MI, USA, 2005; p. 10.
38. Zhao, Y.; Duan, Y.; Zhu, L.; Wang, Y.; Jin, Z. Characterization of coarse aggregate morphology and its effect on rheological and mechanical properties of fresh concrete. *Constr. Build. Mater.* **2021**, *286*, 122940. [CrossRef]
39. Leung, C.K.; Lai, R.; Lee, A.Y. Properties of weted fiber reinforced shotcrete and fiber reinforced concrete with similar composition. *Cem. Concr. Res.* **2005**, *35*, 788–795. [CrossRef]
40. Choi, P.; Yun, K.-K.; Yeon, J.H. Effects of mineral admixtures and steel fiber on rheology, strength, and chloride ion penetration resistance characteristics of wet-mix shotcrete mixtures containing crushed aggregates. *Constr. Build. Mater.* **2017**, *142*, 376–384. [CrossRef]

41. Zhang, Y.; Pan, D.; Qu, X.; Liang, P. Secondary Catalytic Effect of Circulating Ash on the Primary Volatiles from Slow and Fast Pyrolysis of Coal. *Energy Fuels* **2018**, *32*, 1328–1335. [CrossRef]
42. Banfill, P.F.G. The rheology of fresh cement and concrete-a review. In Proceedings of the 11th International Congress on the Chemistry of Cement, Durban, South Africa, 11–16 May 2003.
43. Paiva, H.; Velosa, A.; Veiga, R.; Ferreira, V. Effect of maturation time on the fresh and hardened properties of an air lime mortar. *Cem. Concr. Res.* **2010**, *40*, 447–451. [CrossRef]
44. Greim, M.; Kusterle, W. Rheological measurement of building materials. *Appl. Rheol.* **2004**, *14*, 148–150. [CrossRef]
45. Kostrzanowska-Siedlarz, A.; Gołaszewski, J. Rheological properties of High Performance Self-Compacting Concrete: Effects of composition and time. *Constr. Build. Mater.* **2016**, *115*, 705–715. [CrossRef]
46. Paiva, H.; Velosa, A.; Cachim, P.; Ferreira, V. Effect of metakaolin dispersion on the fresh and hardened state properties of concrete. *Cem. Concr. Res.* **2012**, *42*, 607–612. [CrossRef]
47. Fan, T.; Zhou, G.; Wang, J. Preparation and characterization of a wetting-agglomeration-based hybrid coal dust suppressant. *Process. Saf. Environ. Prot.* **2018**, *113*, 282–291. [CrossRef]
48. Liu, G.; Chen, L. Development of a New Type of Green Switch Air Entraining Agent for Wet-Mix Shotcrete and Its Engineering Application. *Adv. Mater. Sci. Eng.* **2016**, *2016*, 1–9. [CrossRef]
49. Kostrzanowska-Siedlarz, A.; Gołaszewski, J. Rheological properties and the air content in fresh concrete for self compacting high performance concrete. *Constr. Build. Mater.* **2015**, *94*, 555–564. [CrossRef]
50. Pfeuffer, M.; Kusterle, W. Rheology and rebound behaviour of dry-mix shotcrete. *Cem. Concr. Res.* **2001**, *31*, 1619–1625. [CrossRef]
51. Hu, J.; Wang, K. Effect of coarse aggregate characteristics on concrete rheology. *Constr. Build. Mater.* **2011**, *25*, 1196–1204. [CrossRef]
52. Yardimci, M.Y.; Baradan, B.; Taşdemir, M.A. Effect of fine to coarse aggregate ratio on the rheology and fracture energy of steel fibre reinforced self-compacting concretes. *Sadhana* **2014**, *39*, 1447–1469. [CrossRef]
53. Secrieru, E.; Mechtcherine, V.; Schröfl, C.; Borin, D. Rheological characterisation and prediction of pumpability of strain-hardening cement-based-composites (shcc) with and without addition of superabsorbent polymers (sap) at various temperatures. *Constr. Build. Mater.* **2016**, *112*, 581–594. [CrossRef]
54. Westerholm, M.; Lagerblad, B.; Silfwerbrand, J.; Forssberg, E. Influence of fine aggregate characteristics on the rheological properties of mortars. *Cem. Concr. Compos.* **2008**, *30*, 274–282. [CrossRef]
55. Hengjing, B.A.; Zhang, W. Influence of aggregate on the rheological parameters of high-performance concrete. *Concrete* **2003**, *64*, 7–8.
56. Du, L.; Folliard, K.J. Mechanisms of air entrainment in concrete. *Cem. Concr. Res.* **2005**, *35*, 1463–1471. [CrossRef]
57. Burley, R.; Kennedy, B. An experimental study of air entrainment at a solid/liquid/gas interface. *Chem. Eng. Sci.* **1976**, *31*, 901–911. [CrossRef]
58. Jiao, D.; Shi, C.; Yuan, Q.; An, X.; Liu, Y.; Li, H. Effect of constituents on rheological properties of fresh concrete-A review. *Cem. Concr. Compos.* **2017**, *83*, 146–159. [CrossRef]
59. Harini, M.; Shaalini, G.; Dhinakaran, G. Effect of size and type of fine aggregates on flowability of mortar. *KSCE J. Civ. Eng.* **2011**, *16*, 163–168. [CrossRef]
60. Sivakumar, V.; Kavitha, O.; Arulraj, G.P.; Srisanthi, V. An experimental study on combined effects of glass fiber and Metakaolin on the rheological, mechanical, and durability properties of self-compacting concrete. *Appl. Clay Sci.* **2017**, *147*, 123–127. [CrossRef]
61. Tattersall, G.H.; Banfill, P.F. *The Rheology of Fresh Concrete*; Pitman: London, UK, 1983.
62. Struble, L.J. Effects of Air Entrainment on Rheology. *Aci. Mater. J.* **2004**, *101*, 448–456.
63. Rahman, M.A. Effect of geometry, gap, and surface friction of test accessory on measured rheological properties of cement paste. *ACI Mater. J.* **2003**, *100*, 331–339.
64. Kwan, A.; Fung, W. Effects of SP on flowability and cohesiveness of cement-sand mortar. *Constr. Build. Mater.* **2013**, *48*, 1050–1057. [CrossRef]
65. Perrot, A.; Lecompte, T.; Khelifi, H.; Brumaud, C.; Hot, J.; Roussel, N. Yield stress and bleeding of fresh cement pastes. *Cem. Concr. Res.* **2012**, *42*, 937–944. [CrossRef]
66. Laskar, A.I.; Talukdar, S. Rheological behavior of high performance concrete with mineral admixtures and their blending. *Constr. Build. Mater.* **2008**, *22*, 2345–2354. [CrossRef]
67. Beycioğlu, A.; Aruntaş, H.Y. Workability and mechanical properties of self-compacting concretes containing LLFA, GBFS and MC. *Constr. Build. Mater.* **2014**, *73*, 626–635. [CrossRef]
68. Ahari, R.S.; Erdem, T.K.; Ramyar, K. Thixotropy and structural breakdown properties of self consolidating concrete containing various supplementary cementitious materials. *Cem. Concr. Compos.* **2015**, *59*, 26–37. [CrossRef]
69. Vance, K.; Kumar, A.; Sant, G.; Neithalath, N. The rheological properties of ternary binders containing Portland cement, limestone, and metakaolin or fly ash. *Cem. Concr. Res.* **2013**, *52*, 196–207. [CrossRef]
70. Biskri, Y.; Achoura, D.; Chelghoum, N.; Mouret, M. Mechanical and durability characteristics of High Performance Concrete containing steel slag and crystalized slag as aggregates. *Constr. Build. Mater.* **2017**, *150*, 167–178. [CrossRef]

71.   AzariJafari, H.; Tajadini, A.; Rahimi, M.; Berenjian, J. Reducing variations in the test results of self-consolidating lightweight concrete by incorporating pozzolanic materials. *Constr. Build. Mater.* **2018**, *166*, 889–897. [CrossRef]
72.   Yammine, J.; Chaouche, M.; Guerinet, M.; Moranville, M.; Roussel, N. From ordinary rhelogy concrete to self compacting concrete: A transition between frictional and hydrodynamic interactions. *Cem. Concr. Res.* **2008**, *38*, 890–896. [CrossRef]

Article

Load Transfer Behavior and Failure Mechanism of Bird's Nest Anchor Cable Anchoring Structure

Changxing Zhu *, Weihao Zhao and Xu Liu

School of Civil Engineering, Henan Polytechnic University, Jiaozuo 454003, China; zwh@home.hpu.edu.cn (W.Z.); liuxu1997@home.hpu.edu.cn (X.L.)
* Correspondence: zcx@hpu.edu.cn; Tel.: +86-159-3817-1252

Abstract: To research the internal load transfer behavior and failure mechanism of a bird's nest anchor cable anchoring structure based on a pull-out test, a bond-slip failure model is established on the basis of statistical damage theory, and the distribution formula of shear stress at anchorage agent–rock interface is deduced. Combined with theoretical analysis, bird's nest anchor cable pulling out test and particle flow code (PFC) numerical simulation test, as well as axial force distribution of the cable and shear stress distribution of its interface, help reveal its load transfer behavior and failure mechanism. Results show that: (1) The established bond-slip model can reflect the internal load transfer behavior and failure process of bird's nest anchor cable anchorage structure. (2) The shear stress of the anchorage agent interface increases exponentially to the peak value and then decreases exponentially to the residual strength. The process is repeated at every location of the anchorage agent interface. The curve of the axial force and shear stress of the bird's nest anchor cable is a negative exponential distribution with anchorage depth, and the maximum value occurs at the load end. (3) The crack of the anchorage agent interface extends from the load end to the other end and finally cuts through the whole interface. Rock mass generates radial cracks by the split effects of the bird's nest. The failure mode is a combination of the debonding slip of the interface and the shear failure of the rock mass. The shear stress distribution and failure mode of the anchor structure are basically consistent according to laboratory tests and simulation tests, and PFC2D better reflects the internal load transfer behavior, failure mechanism, and failure process of the bird's nest anchor cable under tensile loads.

Keywords: bird's nest anchor cable; pulling-out test; failure mechanism; PFC; fracture process

Citation: Zhu, C.; Zhao, W.; Liu, X. Load Transfer Behavior and Failure Mechanism of Bird's Nest Anchor Cable Anchoring Structure. *Appl. Sci.* **2022**, *12*, 6992. https://doi.org/10.3390/app12146992

Academic Editors: Victor M. Ferreira and Jacek Tomków

Received: 9 June 2022
Accepted: 6 July 2022
Published: 11 July 2022

Publisher's Note: MDPI stays neutral with regard to jurisdictional claims in published maps and institutional affiliations.

1. Introduction

Anchor cables are a common technical means for the reinforcement of deep roadways that has been widely used in many fields, such as underground mine roadways, open-pit slopes, and foundation pits [1–3]. With the increase in mining depth, complex and variable geological structures and stress environment result in a separated roof, roof collapse accidents, and a high repair rate of anchor cables. The ejection phenomenon of broken anchor cables is common [4,5]. In addition to the influence of load and sudden external factors, the anchoring structure of anchor cables is also subject to damage accumulation, which results in bond failure, pull or shear failure of anchor parts, and other forms of anchorage system failure [6–9]. Traditional anchor cable design methods have limitations in deep roadway support [10,11]. Usually, only the axial bearing capacity of the anchor cable is considered, and the interaction of cable-surrounding rock is ignored. Therefore, a reasonable anchor cable design is of considerable significance in controlling the deformation of surrounding rock and improve safety [12,13].

At present, scholars at home and abroad mainly focus on the mechanical effect of anchor cables but ignore the influence of surrounding rock and the relative relationship between the anchor cable and surrounding rock [14]. A large number of studies show that the deformation and failure of surrounding rocks in deep roadways are gradual [15–19]. Previous studies do not reflect the influence of the progressive failure behavior of surrounding rock and mechanical characteristics of cable wells, and the internal load transfer behavior and mechanism of anchor cable still cannot be explained quantitatively and evaluated correctly. Ordinary anchor cables are not adopted to high-deformation environments of surrounding rock. In view of the limitation of traditional anchor cables with respect to mechanical properties and severe engineering environments, new anchor cables with extraordinary mechanical properties should be developed. Existing NPR anchor cables have solved the deformation problem of deep, soft rock roadways [20,21]. The deformation capacity increases to 3000–4000 mm, and the maximum constant resistance is 567.7–800 kN [22]. Some scholars have studied the failure mechanism of anchor cable systems using numerical software, achieving remarkable results. Through theoretical analysis and simulation verification, the mechanical properties of anchor cables under tension–torsion coupling action have been explored, and the tension–torsion coupling coefficient has been proposed and determined [23]. The surrounding rock characteristics of steel arch supports and NPR anchor cable supports are compared with 3DEC software [24].

Bird's nest anchor cables are a new type of anchor cable. Compared with ordinary anchor cables, they has higher anchor force and elongation. The synergistic principle and evaluation method of rock surrounding bird's nest anchor cables are still in the research stage, and the failure process and the mechanisms of anchor structures are not clearly understood. The influence of different anchor cable supporting parameters on the stress diffusion of surrounding rock is studied by FLAC3D, and optimal parameters are proposed [25]. An anchor cable and C tube withstand transverse shear force are developed, and the new scheme can effectively control surface convergence and plastic zone expansion by FLAC3D [26]. Lagging theoretical research results in some accidents during construction or after completion [27,28]. Designers do not have a clear understanding of its action mechanism to deal with the design and application of bird's nest anchor cables. The failure process of rock surrounding bird's nest anchor cables is analyzed and compared with theoretical analysis as well as pull-out tests and PFC simulation tests. Furthermore, the crack evolution process and failure mechanism of bird's nest anchor cables and rock mass, which have certain theoretical significance and practical value in terms of guiding engineering practice, are revealed.

2. Pull-out Test of Bird's Nest Anchor Cables

2.1. Production of Anchorage Structure

(1) Material selection

To study the deformation and failure characteristics of bird's nest anchor cable anchoring structures, a bird's nest with 1×7 strands 17.8 mm, 1720 steel strands, 1200 mm length, 300 mm spacing, and 17.8 mm diameter anchor cable is selected. The specifications and mechanical property parameters of the bird's nest anchor cable are shown in Table 1. The anchorage agent is a MSZ2835 medium-speed resin with a diameter of 28 mm and a single-reel length of 350 mm. To eliminate the size effect, polyvinyl chloride (PVC) pipe is used to anchor samples of different sizes, including 1250 mm height, with 110, 200, and 250 mm outer diameters. The center hole is 1500 mm high, and the outer diameter is 32 mm.

Table 1. Specifications and mechanical property parameters of the bird's nest anchor cable.

Diameter /mm	Strength Level /MPa	Cross-Section Area /mm^2	Perimeter /mm	Fracture Force /kN	Quality /kg·m^{-1}	1% Elongation Load /kN
17.8	1720	191	63	240.2	1.5	294

(2) Preparation of anchor structure

The ratio and strength of similar materials are determined by multiple ratio tests (Table 2). The test material includes 32.5 Portland cement and 0–15 mm well-graded river sand, and the rock mass is made with M15 mortar with a mixture ratio of cement, sand, and water of 1:4.7:1. Strain gauges are pasted inside the anchor cable and rock to monitor the stress–strain of the anchor cable and rock. The rock mass is poured three times, and the strain gauges (BE120-03AA type) are pasted and marked in the designated positions.

Table 2. Ratio and strength of similar materials.

Amount/mm^3			Density	Mean Fraction of Loss Intensity
Portland cement/kg	River sand/kg	Water/kg	2000 kg/m^3	18 MPa
330	1550	330		

The self-made measuring force bolt is prepared before performing the pull-out test. Six pairs of strain gauges are slotted symmetrically on both sides of the bolt cable with 150 mm equal spacing. To avoid the destruction of strain gauges during the process of rotating installation and tension, the strain gauges and wire joints are sealed with 704 silicone rubber and then covered with EP30 Osborne AB glue. The well-tested measuring force bolt is shown in Figure 1.

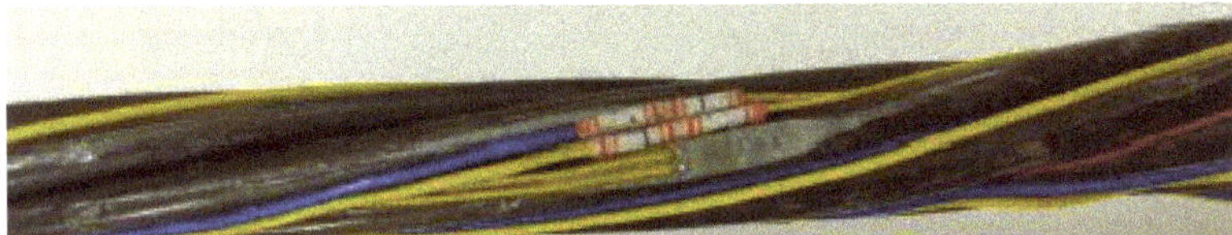

Figure 1. Measuring point figure of bird's nest anchor cable.

The specific preparation is shown in Figure 2, as follows: ① Initial pouring. After placing the 32 mm PVC pipe in the center of the model, the allocated mortar is poured into a PVC pipe with a 110 mm outer diameter and vibrated and poured by plug-in vibrator. The cast similar material model is left for 7 days and then demolded and cured for approximately 7 days. The strain gauge is pasted and marked on the sample surface. ② Second and third castings. After the surface is roughened, a tube with an outer diameter of 200 mm PVC is set on the outside of the poured model, and the above casting process is repeated. The second and third casting processes are then repeated. ③ Cover the outer tube. ④ Installing the anchor agent. The anchor agents are sent into the reserved hole, and the model is finished.

2.2. Test Device

The load equipment is an RRH-6010 double-acting hollow jack with 0.75 kW electric pump power and a 257 mm span. It is composed of a monitoring device, as well as data acquisition and processing system (Figure 3), which can realize real-time data acquisition, dynamic display, real-time chart drawing, and other functions.

A 22-channel analysis system of DH3818N-2 static signal test is used to collect the strain gauge data of the anchor cable and rock sample. Other monitoring devices are composed of a BLR-1/50T pressure sensor and BL100-V-1000 displacement sensor.

2.3. Analysis of Pull-Out Test Results

(1)   Shear stress and displacement curve

When the anchorage agent length is short, the shear stress on the anchorage interface can be considered uniform; that is, the shear stress at any point of the anchorage interface is

$$\tau = \frac{Pdl}{\pi} \tag{1}$$

Figure 4 shows research results [29–31] based on shear stress (τ) and relative shear displacement (ω).

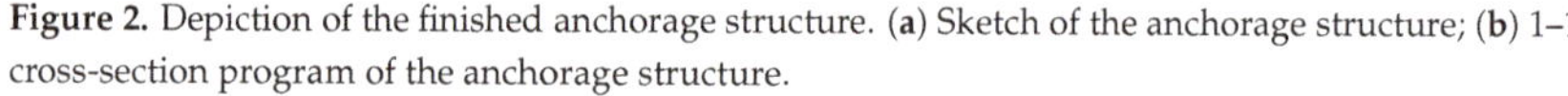

Figure 2. Depiction of the finished anchorage structure. (**a**) Sketch of the anchorage structure; (**b**) 1–1 cross-section program of the anchorage structure.

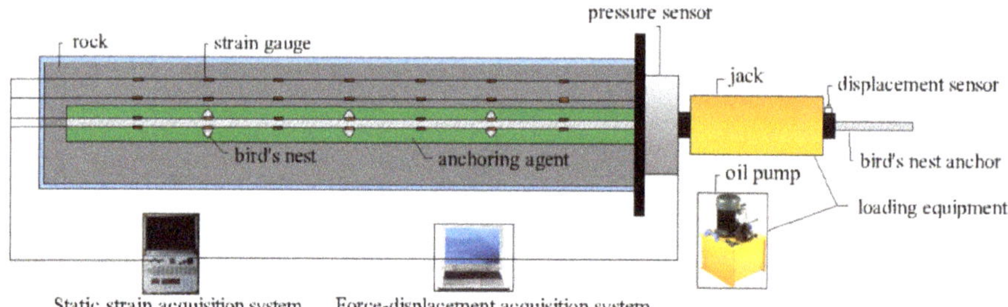

Figure 3. Pulling-out cable test system.

When the ascending stress value reaches peak stress, a linear softening curve occurs until debonding is initiated, followed by a horizontal line that represents the residual strength after debonding is completed (Figure 5).

According to [32,33], the statistical damage constitutive model of the anchorage structure based on Weibull distribution and the Weibull distribution parameter are expressed as follows:

$$\tau = k_1 \omega \exp\left(-k_2 \left(\frac{\omega}{\omega_1}\right)^{k_3}\right) + \beta \tau_p,$$

where ω_1 is displacement, corresponding to peak strain; τ_p is the peak stress; and β is the influence coefficient of residual strength (Figure 5a). $\omega < \omega_1$ is referred to as the nonlinear elastic stage. When $\omega_1 \leq \omega < \omega_2$, the anchorage agent separates from the rib of the bolt, which is referred to as the debonding slip stage. The interfacial adhesion and interlock force disappear simultaneously in the debonding segment, whereas the interfacial friction force gradually approximates the residual strength. When $\omega > \omega_2$, the frictional force depends on the friction surface of the anchorage agent. The period when the shear stress is equal to the residual shear strength of the cable and debonding appears is referred to as the debonding stage. Before the debonding of the model interface, the bond–slip curve is linear, and no damage occurs, so $\beta = 0$ (k_1, k_2, k_3, and β are some parameters of the formula derivation that can be obtained by fitting the experimental data (Figure 5b). The parameters of the ascending stages are 120.9908, 0.2680, 1.25, and 0; the parameters of the descending stages are 0.2213, -4.184, -0.2465, and 0; and the parameters of the horizontal stages are 0, 0, 0, and 0.195.

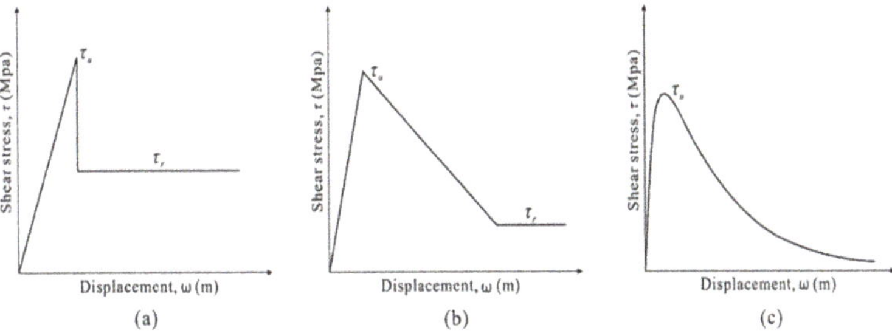

Figure 4. Bond-slip models. (**a**) Model by Cai et al. (**b**) Model by Ren et al. (**c**) Model by Ma et al. τ_u and τ_r are the peak shear strength and residual shear strength, respectively, of a rock bolt.

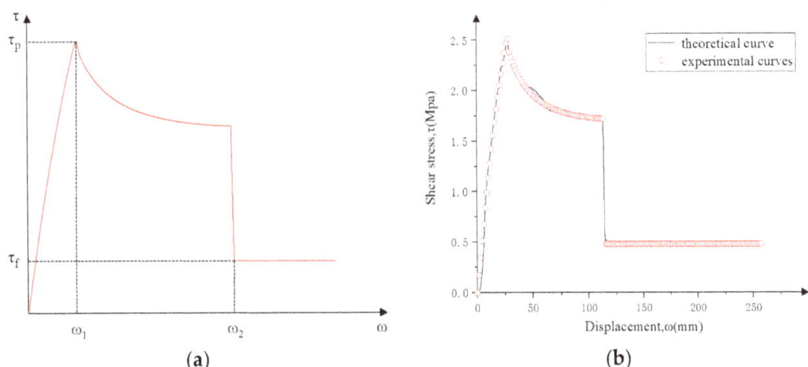

Figure 5. Shear stress–displacement curve of the proposed model. (**a**) Bond–slip models of the proposed model. (**b**) Fitting curve, as well as experimental and theoretical curves.

(2) Distribution law of anchor cable axial force

The obtained axial force distribution curve of the anchor cable under ultimate load (148 kN) is shown in Figure 6. The axial force of the anchor cable gradually decreases along the load end. The peak value is148 kN, and the minimum value is 28 kN (Figure 6).

(3) Distribution law of shear stress at the anchoring–rock interface

The obtained shear stress distribution curves of the anchor cable under three loads are shown in Figure 7. The shear stress of the anchor cable gradually decreases along the anchorage depth. The ultimate shear stress value is 2.37 MPa, and the minimum value tends toward 0.When the axial force is 60 kN, shear stress decreases gradually along the anchorage depth. When the axial force is 100 kN, the shear stress increase first and then decreases. When the axial force is 148 kN, every point of the anchorage agent interface goes through an elastic deformation stage, a debonding slip deformation stage, and a residual deformation stage.

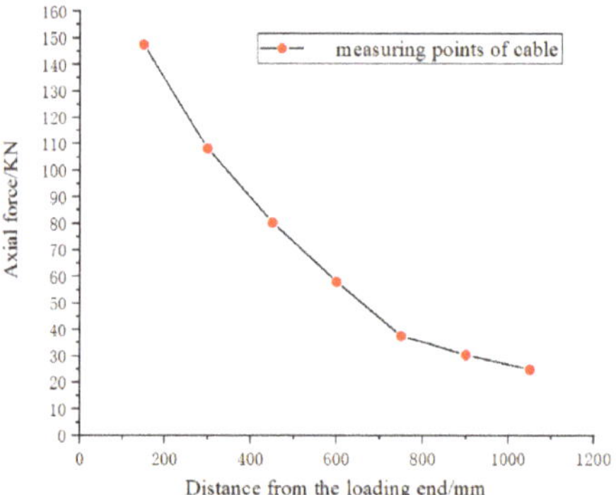

Figure 6. Axial force distribution of the cable.

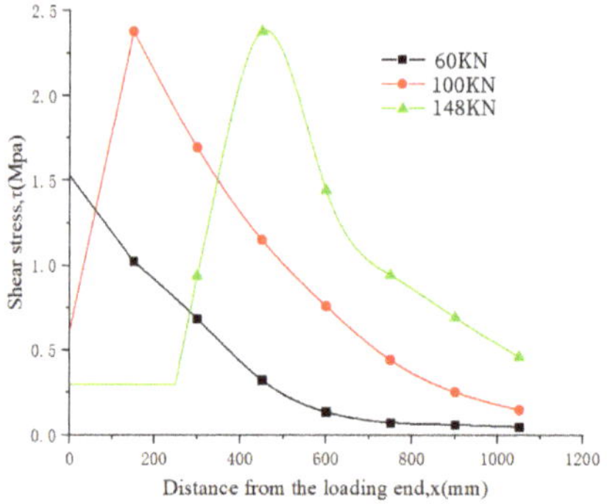

Figure 7. Shear stress distribution law of measuring point at the anchorage agent interface.

(4) Distribution law of shear stress at the rock mass

The distribution law of shear stress in the rock mass when the peak value is 148 kN is shown in Figure 8 with a comparison of the two sets of data. The shear stress near the cable are higher than those far away in the rock mass, and the shear stress of the load end is higher than that of the other end, and the shear stress values of the two interfaces decrease simultaneously.

(5) Fracture mode

Figure 9 shows the final failure mode of the anchoring structure. The cable does not break under tensile load, but the anchorage agent and cable slip at the interface of the rock mass and the anchorage agent. Due to the strong adhesion and shear strength between cable and anchor agent and the mechanical interlocking between bird's nest and anchor agent, it is difficult for relative slip to occur. When the shear stress at the load end is greater than the shear strength at the rock mass anchorage agent interface, internal surface of the rock mass exhibits a radial extension crack caused by the bird's nest dilatation effect and anchorage agent extrusion, and the brittleness of similar material influences the spread range of the crack. The rock mass and anchorage agent interface is the most sensitive and vulnerable part of the whole anchorage system because of the similar material properties.

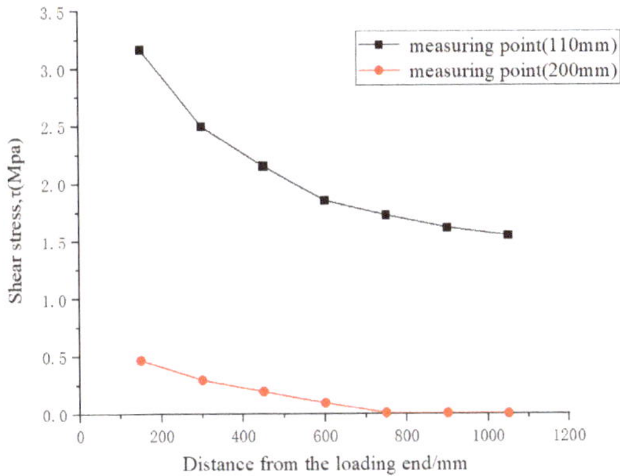

Figure 8. Shear stress distribution of the rock sample.

(a)

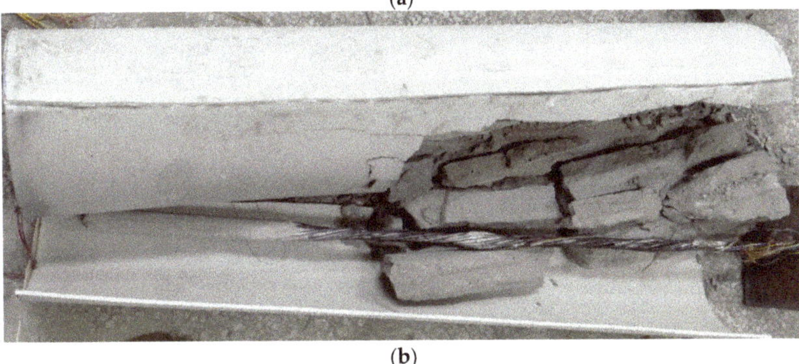

(b)

Figure 9. Failure of the anchoring structure. (a) Cross section; (b) Longitudinal section.

3. PFC Numerical Simulation Test

3.1. PFC Model

The model of the bird's nest anchor cable is established by PFC2D.The model is a cylinder with a base diameter of 250 mm and length of 1200 mm. It is composed of 115,245 particles, including 103,567 rock mass particles and 11,540 anchor agent particles.

Moreover, the rock mass and anchorage agent are represented as gray and blue particles, respectively. The anchorage agent and rock mass are composed of particles with a radius of 0.5–0.85 mm. The cable is composed of 138 particles with a diameter of 17.8 mm and 3 particles with a diameter of 20 mm. According to the measuring point locations, three measuring circles are arranged in the anchorage agent to record the shear stress distribution, and four measuring circles are arranged on the anchorage agent interface to record the shear stress distribution of the anchorage agent interface (Figure 10).

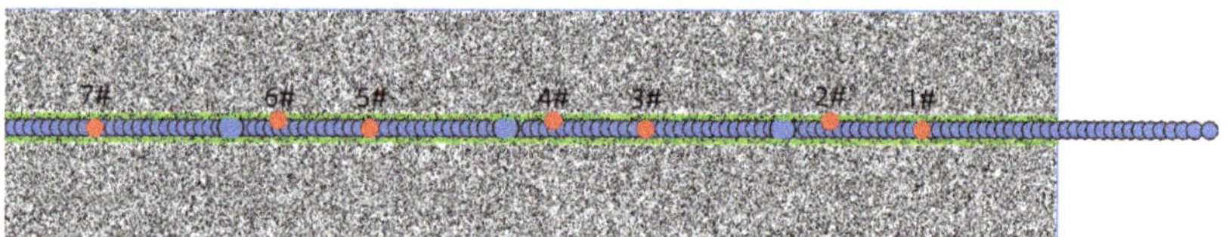

Figure 10. Model and measuring point of PFC simulation test.

To highlight crack, tension and shear cracks are presented as green and red, respectively. The tensile strength and bond force of the rock mass anchored by the PFC are obtained by a uniaxial compression test and Brazilian splitting test and have a value of 3.5 MPa and 3.0 MPa, respectively. The physical and mechanical parameters of the resin anchorage agent and anchor cable are provided by the manufacturer and are equal to 3.0 MPa and 2.5 MPa, respectively. The parameters are constantly adjusted to calibrate the mechanical parameters of similar sample materials. Table 3 shows the parameters of similar materials used in PFC2D.

Table 3. Main parameters of PFC2D in the pull-out test.

Particle Parameter	Numerical Value	Parallel Bond Model Parameters	Numerical Value
Particle radius (mm)	0.5–0.85	Parallel bond modulus	1.4×10^9
Density (kg/m^3)	1.4–1.9×10^3	Normal and tangential stiffness ratio	2.0
Contact modulus between particles	1.2×10^9	Normal critical damping ratio	0.5
Normal and tangential stiffness ratios	1.0	Normal tensile strength	5.1×10^6
Coefficient of friction	0.58	Cohesive force	3.1×10^6

3.2. Failure Process and Failure Mod

Figures 11 and 12 show the axial force distribution law and shear stress evolution process of measuring point by PFC simulation. The results of PFC simulation are consistent with the pull-out test result, as shown in Figures 5–7, 11 and 12. Steps 2600 and 10,000 are the boundary lines of the elastic deformation stage and bird's nest split rock failure stage, respectively.

Combined with the pull-out test result and PFC simulation result, the crack evolution processes are divided into three stages, namely the elastic deformation stage, the debonding slip stage, and the split failure of the bird's nest, as shown in Figures 12–14. The failure process is explained as follows:

Cracks can be divided into three stages, as shown in Figure 14. Stage I (elastic deformation stage): before step 1350, shear crack dominates the evolution process of cracks. In steps 1350–2600, tension cracks exceed shear cracks. The curve of this stage rises, and the shear stress–slip curve is in an elastic state. The bond between the anchor cable and anchor agent mainly consists of chemical bond force, friction force, and mechanical interlocking force. Due to the relatively small tensile load, the anchorage agent is the elastic coordinated deformation stage. When the shear stress of interface reaches shear stress strength, the anchor cable moves the small slip at the load end, which results in the gradual loss of chemical bond force of the anchorage agent, and the anchor agent interface exhibits a few cracks. Before cracks develop near the bird's nest, cracks extend inwards with anchorage agent depth, but the rock mass does not fail. After cracks develop near the bird's nest, they continue to extend inwards, and rock mass failure occurs.

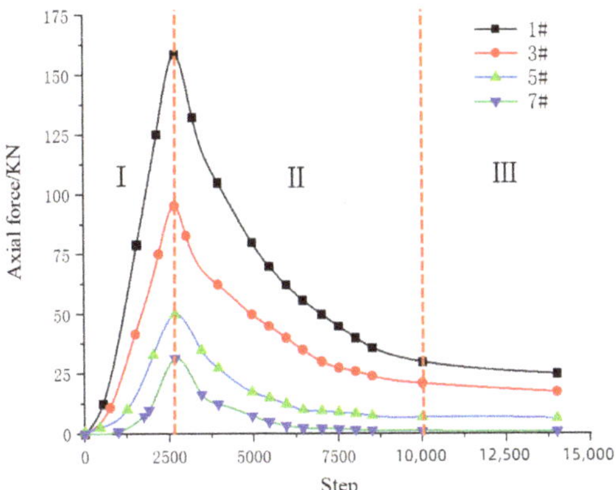

Figure 11. Axial force distribution law of measuring point.

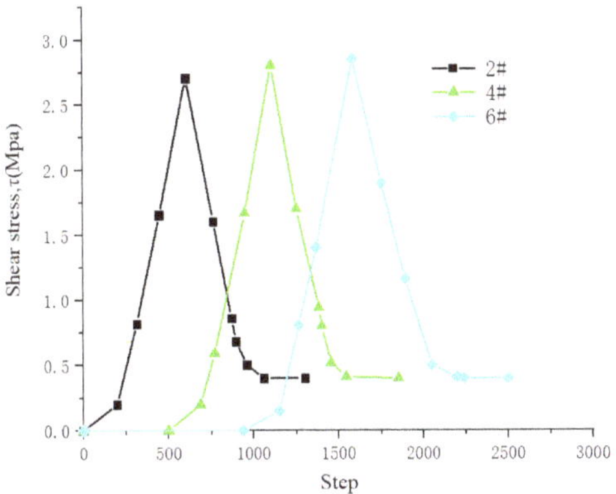

Figure 12. Shear stress evolution process of measuring point.

Stage II (debonding slip stage): loading occurs at steps 2600–10,000, the tension crack effect dominates, and shear crack number has a constant value. With additional debonding length, the tensile load of the anchor cable decreases. The relative slip of anchoring interface is no longer coordinated with the increase in load. The shear strength of the bird's nest anchor cable is composed of adhesion, mechanical interlocking, and friction in the axial direction at the coupling interface between the anchor cable and the anchorage agent. As the load increases, interfacial slip and shear dilation appear, and in turn, the interfacial radial force and friction force, as well as the peak shear strength, are elevated. A debonding region occurs at the load end and gradually expands to the other end.

Stage III (bird's nest split rock failure stage): After step 10,000, the shear crack and tension crack number are basically a constant value, and a few cracks extend along the rock mass depth. The tensile load of the anchor cable declines significantly up to the residual tensile stress. The friction force and the extruding force of the bird's nest hinder the anchor cable from being pulled out. At this stage, the interface of the bird's nest generates radial cracks, and cracks rapidly expand the rock mass. The radial stress produced by mechanical biting force balances the annular tensile stress of the anchor agent, with only friction resistance remaining. The anchorage agent is sheared, and rock crumbs are drawn unceasingly with the pull-out of the anchor cable.

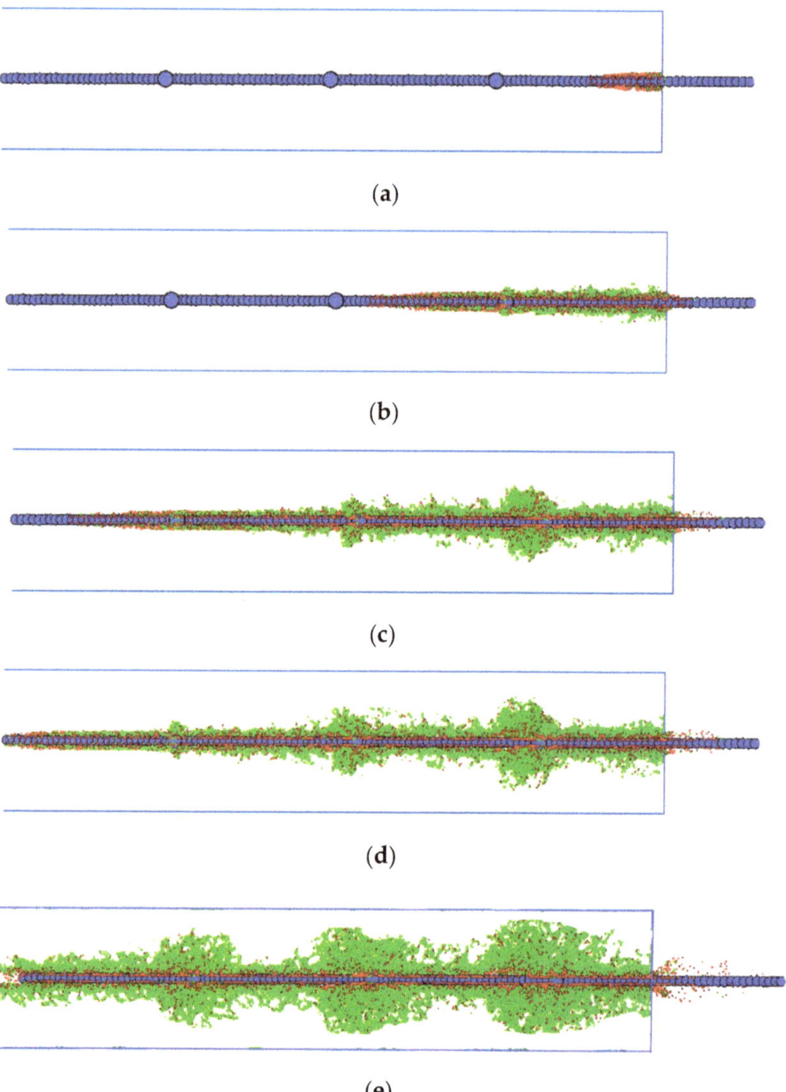

Figure 13. Failure process of the anchoring structure during PFC simulation test. (**a**) Step 250 (crack number 4000, shear crack 2500, tension crack 1500); (**b**) step 1250 (crack number 28000, shear crack 16000, tension crack 12000); (**c**) step 2400 (crack number 58000, shear crack 19000, tension crack 39000); (**d**) step 2600 (crack number 62000, shear crack 20000, tension crack 42000); (**e**) step 10000 (crack number 87000, shear crack 21000, tension crack 66000).

The failure mode is a mixed failure between debonding slip of the interface and shear failure of the rock mass (Figure 9). PFC2D can better reflect the internal load transfer behavior, failure mechanism, and failure process of the bird's nest anchor cable under tensile loads.

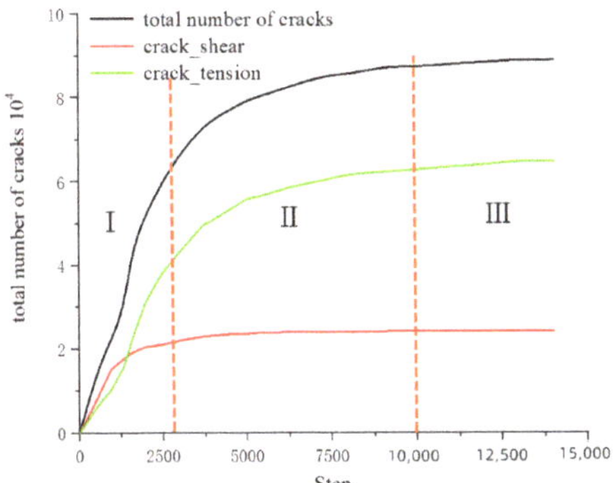

Figure 14. Crack evolution.

4. Conclusions

(1)　A bond-slip model of bird's nest anchor cable is established on the basis of a statistical damage constitutive model. The shear stress and slip curve of the anchorage structure are divided into a nonlinear elastic deformation stage, a debonding slip stage, and a residual stage. With an increase in the pull-out load, the shear stress of the anchorage agent interface increases exponentially to the peak value from the load end, then decreases, and finally stabilizes to the residual strength. The process transmits every point with the anchorage depth. The model parameters are acquired by fitting data, and the theory curve and test curves are similar, thereby verifying the reasonability of the proposed mode.

(2)　The axial force of the anchor cable declines negatively along the load direction; the maximum value (148 kN) is near the load end, and the minimum value (28 kN) is at the other end. The shear stress of the rock mass decreases negatively along the load direction and is transferred from the load end to the other end, but the shear stress value near anchorage cable value is higher than that far away from the anchorage structure.

(3)　The failure of the anchorage structure is divided into three stages stage according to PFC simulation: an elastic deformation stage, a debonding slip stage, and a splitting rock failure. The anchorage structure of the load end first exhibits a small number of cracks (crack number: 4000), then extends along the interface of the anchorage agent (crack number: 62000), and then penetrates into the anchorage agent–rock mass interface (crack number: 87000). The failure mode of the anchoring structure is mixed failure, namely debonding slip of the interface and shear failure of rock mass. PFC can simulate the crack evolution process of the bird's nest anchor cable well.

Author Contributions: Methodology, X.L.; Writing—original draft, W.Z.; Writing—review & editing, C.Z. All authors have read and agreed to the published version of the manuscript.

Funding: The research was funded by the National Natural Sciences Foundation of China (No. 51874119), the Education Department of Henan Province (No.2011A440003), and the Doctor Foundation of Henan Polytechnic University (No.B2009-96).

Institutional Review Board Statement: Not applicable.

Informed Consent Statement: Not applicable.

Conflicts of Interest: The authors declare no conflict of interest.

References

1. Chen, G.; Chen, T.; Chen, Y.; Huang, R.; Liu, M. A new method of predicting the prestress variations in anchored cables with excavation unloading destruction. *Eng. Geol.* **2018**, *241*, 109–120. [CrossRef]
2. Cui, J.; Wang, W.; Yuan, C.; Cao, L.; Guo, Y.; Fan, L. Study on Deformation Mechanism and Supporting Countermeasures of Compound Roofs in Loose and Weak Coal Roadways. *Adv. Civ. Eng.* **2020**, *2020*, 5610058. [CrossRef]

3. Tao, Z.; Xu, H.; Zhu, C.; Lv, Q.; Wang, Y. The Study of the Supernormal Mechanical Properties of Giant NPR Anchor Cables. *Shock Vib.* **2020**, *2020*, 262190. [CrossRef]
4. Wang, X.; Wang, G.; Wu, B.; Liu, S. Study on the Lagging Support Mechanism of Anchor Cable in Coal Roadway Based on FLAC3D Modified Model. *Adv. Civ. Eng.* **2021**, *2021*, 9919454. [CrossRef]
5. Xiang, Y.; Chen, Z.; Yang, Y.; Lin, H.; Zhu, S. Dynamic response analysis for submerged floating tunnel with anchor-cables subjected to sudden cable breakage. *Mar. Struct.* **2018**, *59*, 179–191. [CrossRef]
6. Jiang, Q.; Feng, X.-T.; Cui, J.; Li, S.-J. Failure Mechanism of Unbonded Prestressed Thru-Anchor Cables: In Situ Investigation in Large Underground Caverns. *Rock Mech. Rock Eng.* **2014**, *48*, 873–878. [CrossRef]
7. Wu, J.; Xian, G.; Li, H. A novel anchorage system for CFRP cable: Experimental and numerical investigation. *Compos. Struct.* **2018**, *194*, 555–563. [CrossRef]
8. Liu, X.; Yao, Z.; Xue, W.; Wang, X.; Huang, X. Experimental Study of the Failure Mechanism of the Anchorage Interface under Different Surrounding Rock Strengths and Ambient Temperatures. *Adv. Civ. Eng.* **2021**, *2021*, 6622418. [CrossRef]
9. Kim, J.R.; Kwak, H.-G.; Kim, B.-S.; Kwon, Y.; Bouhjiti, E.M. Finite element analyses and design of post-tensioned anchorage zone in ultra-high-performance concrete beams. *Adv. Struct. Eng.* **2018**, *22*, 323–336. [CrossRef]
10. Wang, J.; Ning, J.-G.; Tan, Y.-L.; Hu, S.-C.; Guo, W.-Y. Deformation and failure laws of roadway surrounding rock and support optimization during shallow-buried multi-seam mining. *Geomat. Nat. Hazards Risk* **2020**, *11*, 191–211. [CrossRef]
11. Tai, Y.; Xia, H.; Meng, X.; Kuang, T. Failure mechanism of the large-section roadway under mined zones in the ultra-thick coal seam and its control technology. *Energy Sci. Eng.* **2019**, *8*, 999–1014. [CrossRef]
12. Shan, R.; Bao, Y.; Huang, P.; Liu, W.; Li, G. Study on Double-Shear Test of Anchor Cable and C-Shaped Tube. *Shock Vib.* **2021**, *2021*, 9948424. [CrossRef]
13. Wang, S.R.; Wang, Z.L.; Chen, Y.B.; Wang, Y.H.; Huang, Q.X. Mechanical Performances Analysis of Tension-Torsion Coupling Anchor Cable. *Int. J. Simul. Model.* **2020**, *19*, 231–242. [CrossRef]
14. Zeng, Y.F.; Wu, Y.P.; Lai, X.P.; Wei, C. Analysis of Roof Caving Instability Mechanism of Large-Section Roadway Under Complex Conditions. *J. Min. Saf. Eng.* **2009**, *26*, 423–427. [CrossRef]
15. Li, S.; Wang, Q.; Wang, H.; Jiang, B.; Wang, D.; Zhang, B.; Li, Y.; Ruan, G. Model test study on surrounding rock deformation and failure mechanisms of deep roadways with thick top coal. *Tunn. Undergr. Space Technol.* **2015**, *47*, 52–63. [CrossRef]
16. Pan, S.; Liu, S.; Cao, L.; Guo, J.; Yuan, C. Deformation Failure and Support Test of Surrounding Rock in Deep Arched Roadway with Straight Wall. *Adv. Civ. Eng.* **2021**, *2021*, 3208974. [CrossRef]
17. Cao, C.; Jan, N.; Ren, T.; Naj, A. A study of rock bolting failure modes. *Int. J. Min. Sci. Technol.* **2013**, *23*, 79–88. [CrossRef]
18. Kang, H.; Wu, Y.; Gao, F.; Lin, J.; Jiang, P. Fracture characteristics in rock bolts in underground coal mine roadways. *Int. J. Rock Mech. Min. Sci.* **2013**, *62*, 105–112. [CrossRef]
19. Li, S.-C.; Wang, H.-T.; Wang, Q.; Jiang, B.; Wang, F.-Q.; Guo, N.-B.; Liu, W.-J.; Ren, Y.-X. Failure mechanism of bolting support and high-strength bolt-grouting technology for deep and soft surrounding rock with high stress. *J. Central South Univ.* **2016**, *23*, 440–448. [CrossRef]
20. Feng, J.; Xu, X.; Liu, P.; Ma, F.; Ma, C.; Tao, Z. Slope Sliding Force Prediction via Belief Rule-Based Inferential Methodology. *Int. J. Comput. Intell. Syst.* **2021**, *14*, 965–977. [CrossRef]
21. Zhang, K.; Yang, X.; Cui, X.; Wang, Y.; Tao, Z. Numerical Simulation Analysis of NPR Anchorage Monitoring of Bedding Rock Landslide in Open-Pit Mine. *Adv. Civ. Eng.* **2020**, *2020*, 8241509. [CrossRef]
22. Tao, Z.; Zhu, C.; He, M.; Karakus, M. A physical modeling-based study on the control mechanisms of Negative Poisson's ratio anchor cable on the stratified toppling deformation of anti-inclined slopes. *Int. J. Rock Mech. Min. Sci.* **2021**, *138*, 104632. [CrossRef]
23. Wang, F.; Guo, Z.; Qiao, X.; Fan, J.; Li, W.; Mi, M.; Tao, Z.; He, M. Large deformation mechanism of thin-layered carbonaceous slate and energy coupling support technology of NPR anchor cable in Minxian Tunnel: A case study. *Tunn. Undergr. Space Technol.* **2021**, *117*, 104151. [CrossRef]
24. Sun, X.; Zhang, B.; Gan, L.; Tao, Z.; Zhao, C. Application of Constant Resistance and Large Deformation Anchor Cable in Soft Rock Highway Tunnel. *Adv. Civ. Eng.* **2019**, *2019*, 4347302. [CrossRef]
25. Li, C.; Zhang, W.; Huo, T.; Yu, R.; Zhao, X.; Luo, M. Failure Analysis of Deep Composite Roof Roadway and Support Optimization of Anchor Cable Parameters. *Geofluids* **2021**, *2021*, 5610058. [CrossRef]
26. Guo, X.; Zhao, Z.; Gao, X.; Ma, Z.; Ma, N. The Criteria of Underground Rock Structure Failure and Its Implication on Rockburst in Roadway: A Numerical Method. *Shock Vib.* **2019**, *2019*, 7509690. [CrossRef]
27. Jing, H.W.; Yin, Q.; Zhu, D.; Sun, Y.J.; Wang, B. Experimental study on the whole process of instability and failure of anchorage structure in surrounding rock of deep-buried roadway. *J. China Coal Soc.* **2020**, *3*, 889–901. [CrossRef]
28. Xuegui, S.; Yanbin, L.; Yongkang, Y. A Research into Extra-thick compound mudstone roof roadway failure mechanism and security control. *Procedia Eng.* **2011**, *26*, 516–523. [CrossRef]
29. Cai, Y.; Esaki, T.; Jiang, Y. An analytical model to predict axial load in grouted rock bolt for soft rock tunnelling. *Tunn. Undergr. Space Technol.* **2004**, *19*, 607–618. [CrossRef]
30. Ren, F.; Yang, Z.; Chen, J.; Chen, W. An analytical analysis of the full-range behaviour of grouted rockbolts based on a tri-linear bond-slip model. *Constr. Build. Mater.* **2010**, *24*, 361–370. [CrossRef]

31. Ma, S.; Nemcik, J.; Aziz, N. An analytical model of fully grouted rock bolts subjected to tensile load. *Constr. Build. Mater.* **2013**, *49*, 519–526. [CrossRef]
32. Song, Y.; Li, Y. Study on the constitutive model of the whole process of macroscale and mesoscale shear damage of prestressed anchored jointed rock. *Bull. Eng. Geol. Environ.* **2021**, *80*, 6093–6106. [CrossRef]
33. Chen, J.; Yang, S.; Zhao, H.; Zhang, J.; He, F.; Yin, S. The Analytical Approach to Evaluate the Load-Displacement Relationship of Rock Bolts. *Adv. Civ. Eng.* **2019**, *2019*, 2678905. [CrossRef]

 *applied sciences*

Review

Mechanical Properties and Microstructure of Shotcrete under High Temperature

Guoming Liu [1,2], Jipeng Zhao [3,*], Zhixue Zhang [4], Chenglong Wang [1,*] and Qianqian Xu [1,*]

[1] College of Safety and Environmental Engineering, Shandong University of Science and Technology, Qingdao 266590, China; skd995978@sdust.edu.cn

[2] State Key Laboratory of Mining Disaster Prevention and Control Co-Founded by Shandong Province and Ministry of Science and Technology, Shandong University of Science and Technology, Qingdao 266590, China

[3] College of Energy and Mining Engineering, Shandong University of Science and Technology, Qingdao 266590, China

[4] Ventilation and Dust Prevention Department of Feicheng Mining Group Co., Ltd., Tai'an 271600, China; zxzhang@gsm.pku.edu.cn

* Correspondence: jipengzhao0429@163.com (J.Z.); wclcm@sdust.edu.cn (C.W.); 13864240150@163.com (Q.X.)

Citation: Liu, G.; Zhao, J.; Zhang, Z.; Wang, C.; Xu, Q. Mechanical Properties and Microstructure of Shotcrete under High Temperature. *Appl. Sci.* **2021**, *11*, 9043. https://doi.org/10.3390/app11199043

Academic Editors: Carlos Thomas and Doo-Yeol Yoo

Received: 30 June 2021
Accepted: 14 September 2021
Published: 28 September 2021

Publisher's Note: MDPI stays neutral with regard to jurisdictional claims in published maps and institutional affiliations.

Abstract: High temperature is recognized as one of the extreme environments in the application of shotcrete which significantly deteriorate the performance of shotcrete. This paper reviews the mechanical properties and microstructure of shotcrete under high temperature. First of all, this paper reviews the cause of formation of high ground temperature. Based on these causes, the author establishes a heat transfer model with a spiral shape by introducing a multidimensional morphological formula into the heat conduction process. Then, the paper reviews the influence of high temperature on the mechanical and micro properties of shotcrete, the cooling technology under high temperature, and the optimization research technology of shotcrete. The author discusses the influence of high temperature on the thermal parameters and the deformation of shotcrete from the perspective of thermodynamics. Multiple studies have shown that the irregular movement and disorderly overlapping of molecules in the shotcrete caused by the high temperature environment result in the premature termination of the hydration reaction of cement in shotcrete. Finally, the author suggests the challenges of high-temperature shotcrete in term of the process structure, performance optimization, and application in special engineering fields. The research in this paper intends to give guidance to those conducting shotcrete research under high temperature, and to promote the further development of shotcrete technology.

Keywords: high temperature; shotcrete; mechanical properties; microscopic properties

1. Introduction

With the swift global development of tunnels, mines, subways, water conservancy and hydropower projects, and so on, shotcrete, as an advanced support method, is widely applied to surrounding rock control and roadway closure. Shotcrete is a kind of concrete formed by mixing concrete materials, such as gel material, aggregate, and so on, into the ejection equipment, by means of compressed air or other power transmission, and sprayed onto the spray surface at high speed [1]. Shotcrete technology was first used in mining and civil engineering by the United States in 1914. It has a history of more than 100 years. This method is used in underground and tunnel support, infrastructure repair and rehabilitation, slope stabilization, and in areas difficult to reach with conventional concrete, such as bridge piers and beam soffits [2]. Shotcrete has the characteristics of high compressive strength, good durability, and wide range of strength grades [3–5]. As a supporting material of roadway, shotcrete can not only prevent the oxidation of surrounding rock, but also plays a supporting role for the roadway. With the development of deep mines and deep tunnels, shotcrete is facing great challenges. Especially deep underground, high temperature accelerates the deterioration of shotcrete [6–9]. According to relevant reports, the rock wall temperature of Sangzhuling Tunnel in the Yarlung Zangbo River Canyon in China reached 89 °C, and it was identified as a class I risk tunnel. The high temperature of 75 °C and the extreme temperature of 170 °C were encountered in the construction of Anfang tunnel and the third hydropower station of Heibu in Japan. The highest original rock temperature is 46.8 °C, at −980 m level in Sanhejian coal mine. The underground rock temperature of the Mbonig gold mine in South Africa reaches 65.6 °C.

Many scholars have shown that there is a certain relationship between temperature and depth, and the rock temperature gradient is about 3–4 °C/100 m [10–12]. However, in a deep environment, due to the different rock properties and the possible existence of large fault zones, the change of temperature with temperature gradient is not obvious. Figure 1 shows the distribution of temperature with depth in deep tunnels and deep mines in some projects. From the figure, it could be concluded that the variation between temperature and depth in underground projects was a nonlinear relationship, that is, the ground temperature in deep environments shows an abnormal pattern [13,14].

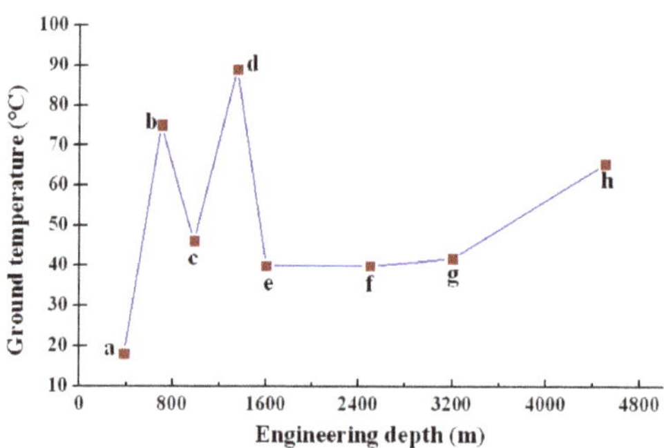

Figure 1. Temperature distribution of tunnels and mines with different depths.

In Figure 1: a—Albert tunnel in Austria; b—Anfang highway tunnel in Japan; c—Sanhejian coal mine in China; d—Sangzhuling tunnel in China; e—Qinling tunnel of Xikang railway in China; f—Franco-Yibulangfeng highway tunnel; g—Kolar gold mine in India; h—Mponig gold mine in South Africa.

When shotcrete as support structure contacts with high-temperature rock wall, the performance of shotcrete may change. High temperature leads to the deterioration of shotcrete structure and the weakening of roadway stability. In recent years, many scholars have studied the mechanical properties and micro-morphology of shotcrete under high temperature. For example, Lee and Yang et al. [15,16] studied the variation of shear properties of shotcrete with granite cementation surface roughness and temperature. The results showed that there were critical values for the effects of granite cementation surface roughness and temperature on shear strength. Moreover, the temperature was the most important factor affecting the shear performance of shotcrete. Wang et al. [17,18] studied the impermeability of shotcrete under standard working conditions and variable temperature conditions. The results showed that the temperature destroys the dense structure inside the shotcrete, resulting in an increase in the impermeability of the shotcrete. Wang and Cui et al. [19] studied the effect of temperature on the bond strength between shotcrete and rock surface, and found that the bond strength first increased and then decreased with the increase of temperature. Dong et al. [20] studied the fracture process of the interface between shotcrete and rock, and found that under high temperature, the structure of shotcrete fracture surface was loose and powder particles increased. Yang et al. [21] studied the internal damage mechanism of shotcrete under high temperature based on CT technology and X-ray, and the results showed that the internal pores of shotcrete present the trend of small holes gradually developing into large pores with the increase of temperature.

Lu et al. [22,23] studied the residual properties of shotcrete after high temperature, and found that the main factors affecting the residual strength were curing conditions and cooling methods. Kjellsen et al. [24] studied the consolidation process of concrete in a thermal environment, and found that the compaction degree of the concrete decreased, and internal cracks occurred during the solidification process. Then, the microstructure of concrete was scanned by scanning electron microscopy (SEM), and it was found that the cracks developed along the interior of the aggregate. Akca et al. [25] studied the structural performance of high-performance shotcrete under high temperature, and concluded that the overall strength of shotcrete showed an upward trend under high temperature.

Although there are many studies on the performance changes of shotcrete under high temperature, there are few systematic reviews on the mechanical and micro properties of shotcrete. Therefore, this article systematically describes the development of mechanical properties and microstructure of shotcrete under high temperature. The research in this paper is intended to give guidance to those conducting shotcrete research under high temperature, and to promote the further development of shotcrete technology.

2. Causes of High Ground Temperature and Its Influence on Shotcrete Structure

In deep mines and deep tunnels, the phenomenon of high ground temperature is the biggest problem in the process of excavation and support. It will not only worsen the construction environment and reduce the labor production efficiency, but also change the shotcrete structure and affect the stability of the tunnel, thus threatening the safety of construction personnel. This chapter will discuss the cause of high ground temperature and the influence of high temperature on shotcrete structure.

2.1. Causes of Formation of High Ground Temperature

The earth's temperature is produced by the decay of radioactive elements in the earth's interior, and heat is accumulated in the earth's crust [26,27]. Then, heat is transmitted through rocks to the earth's surface. Because of the strong gravitational field inside the earth, there is a lot of energy in the inner core of the earth, and the temperature of the core is about 6000 °C. However, most of the energy gradually disappears in the process of transmission. Because of the low thermal conductivity and poor heat transfer performance of rocks, heat energy builds easily in the rock mass, which makes the high temperature phenomenon appear in deep tunnels or mines; the remaining small part of energy will continue to transmit to the earth's surface [28–30]. Therefore, the main heat source for the formation of the earth's surface geothermal field is from the earth's interior. Although the transmission of energy is very small, the geothermal field formed in this process is still very harmful to human infrastructure work. In addition, the undulation and structural form of the base also have a certain influence on the formation temperature. When the tunnels and mines are deep in large fault zones, the temperature difference may reach 2–4 °C/100 m [31]. This is a further challenge to the original high-temperature environment.

The heat of the earth's surface layer is from the earth's interior, and is transmitted to the shallow part of the earth's surface through heat conduction [32]. This mode of transmission is one-dimensional propagation. The temperature change can be expressed by one-dimensional conduction Equation (1):

$$\frac{d^2\theta}{dz^2} = 0 \tag{1}$$

In Equation (1): θ—temperature; Z—depth

However, in a deep environment, due to the complex geological environment, most of the strata are in a non-horizontal state. The transmission mode of temperature is diverse and controlled by many factors. At this time, the heat conduction occurs in a two dimensional or three dimensional shape. The one-dimensional conduction formula in the above Equation (1) cannot reflect the temperature propagation law at this time. Therefore, the following Equation (2) is introduced [33]:

$$\frac{\partial}{\partial x}\left(\frac{\partial \theta}{\partial x}\right) + \frac{\partial}{\partial z}\left(\lambda \frac{\partial \theta}{\partial z}\right) + A = 0 \tag{2}$$

In Equation (2): λ—thermal conductivity of rock; A—heat production of radioactive elements in rock.

The above Equation (2) shows that depth, rock thermal conductivity, and radioactive element heat production are the main factors affecting the temperature in deep environments. This is because the greater the depth, the greater the gravitational potential energy and the more heat generated. In a deep environment, the closer to the crust, the greater the influence of radioactive element decay heat production. However, rock is an anisotropic heterogeneous material, and its thermal conductivity is variable and generally decreases with the increase of temperature. Therefore, in a deep environment, due to the low thermal conductivity of rock, it shows strong heat storage capacity, resulting in a high-temperature environment.

Heat can be imagined as a fluid, and the fluid can be divided into steady-state type and non-stationary type in the process of transmission [34]. The heat transfer process is an unsteady state. Figure 2 shows the physical structure model of heat transfer. Luo et al. [35,36] analyzed the heat transfer phenomenon in the flow process, and results showed that heat flow is a process of energy exchange. Therefore, the author divides the heat transfer process into three types: linear, curved, and spiral. Due to the thermal conductivity of rock, the heat transferred from the deep to the ground is mainly linear [37,38]. In the process of tunnel construction, the heat is mainly curved and spiral; the heat flow transferred by curve will continue to diffuse around; the heat transferred in this form is not stored much in the rock. The heat transferred by spiral type will not disappear, but will accumulate continuously in the process of transmission, which will enhance the heat storage capacity of rock. At the same time, the spiral transmission is also the dominant part of deep high temperatures.

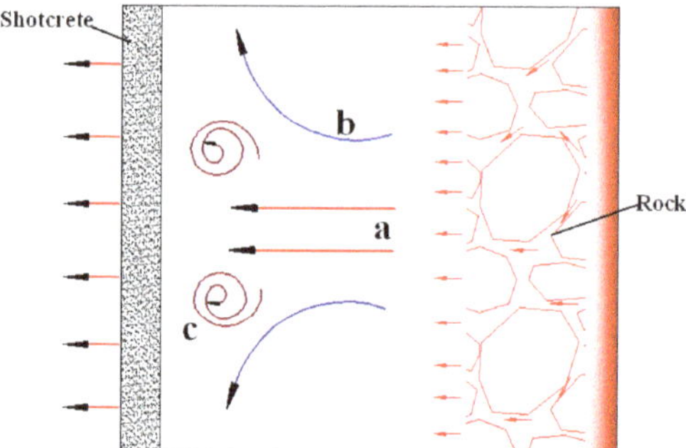

Figure 2. Heat transfer model: (a) linear type, (b) curve type, (c) spiral type.

2.2. Influence of High Temperature on Shotcrete Structure

In a deep construction environment, the heat flow transmission path in the stratum is cut off indirectly because of the tunnel excavation. As a result, the heat storage capacity of the roadway is strengthened. Combined with the impact pressure in the formation, it can easily cause structural instability and structural deformation of the roadway. On the one hand, the shotcrete layer may fall off seriously; on the other hand, it will affect the selection of construction materials and the durability of shotcrete [39,40]. Furthermore, the additional temperature stress can also cause the shotcrete lining to crack, which will have a great impact on the initial support of the roadway. For example, Zhu et al. [41] introduced a thermal crack mechanism and considered that shotcrete was a heterogeneous mixed material, and the thermodynamic properties of shotcrete aggregate and cement paste were not matched under the influence of temperature. As a result, radial and circumferential cracking will occur inside the shotcrete. The internal cracking of aggregate and cement paste will occur, resulting in the spalling of shotcrete surface, as shown in Figure 3.

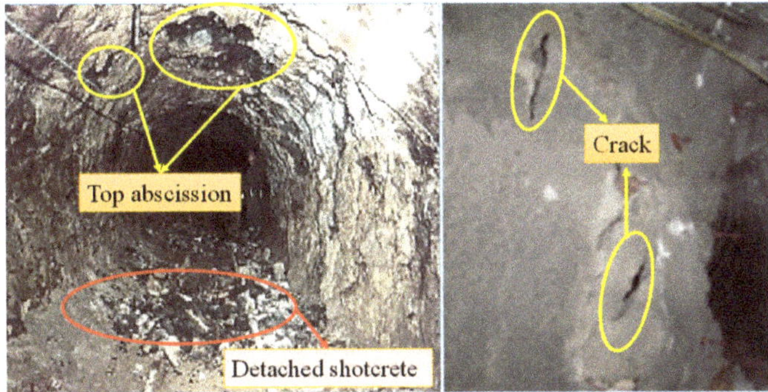

Figure 3. Shotcrete layer falling off at high temperature.

In order to further study the mechanism of cracks in roadway concrete under high temperature, we summarize the early cracking phenomenon of shotcrete. The lining shotcrete under a high-temperature heat-damaged roadway can easily deteriorate in the process of pouring and curing. One reason is that the high temperature makes the water in the fresh concrete evaporate rapidly, forming large voids or cracks; another reason is that when shotcrete adheres to high-temperature surrounding rock, it has high-temperature stress. In addition, high temperature changes the hydration products and progress of cement. These reasons are worthy of in-depth study. For example, Li et al. [42] studied the influence of temperature, water/cement ratio, and fly ash content on the early crack resistance of high-performance shotcrete, and results indicated that temperature was the primary factor affecting the early crack-resistance performance. Huang et al. [43] studied the performance of early-age concrete under fire and found that the residual compressive strength of concrete showed a downward trend with temperature. It was also found that the internal damage to concrete specimens was serious: the hydration products C-S-H gel and Ca $(OH)_2$ crystals were basically decomposed, leading to an imbalance of concrete structure stability. Qin et al. [44] studied high-strength shotcrete at 80 °C. They found that the early performance of shotcrete was significantly improved, and the frost resistance and chloride ion penetration resistance of shotcrete were improved. Some scholars have carried out X-ray diffraction (XRD) tests on shotcrete at different temperatures. The XRD spectra of shotcrete at 30 °C and 60 °C is shown in Figure 4; results show that the content of CH crystal at 60 °C is higher than at 30 °C. This shows that the hydration degree of shotcrete cement is better at 60 °C, and more C-S-H gel is generated. Therefore, obtaining a certain temperature is conducive to the early strength development of shotcrete.

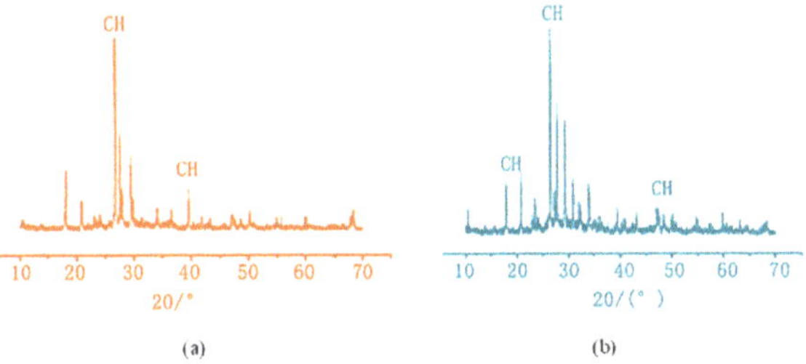

Figure 4. XRD patterns of shotcrete at different temperatures: (**a**) 30 °C, (**b**) 60 °C [45].

Hiremath et al. [46] studied the development of early strength of shotcrete under the conditions of hot water bath and hot air curing; they found that the early strength of shotcrete showed an upward trend with the increase of temperature and that hot water bath curing was conducive to the development of early strength of shotcrete. Wonsuk et al. [47] studied slag shotcrete and found that under high-temperature conditions, the output of CO_2 in cement decreased and the early strength of shotcrete was significantly improved. The linear expansion value of slag shotcrete is very low and the pozzolanic activity is high, which is conducive to the development of shotcrete strength at the later stage [48]. As shown in Figure 5, D'Aloia et al. [49] carried out a numerical simulation on the cracking performance of tunnel lining shotcrete. The results showed that early creep has a good effect (Figure 5a). When the creep is ignored, the thermal damage occurs in a large range of shotcrete lining structure (Figure 5b). By comparing the simulation results of thermal shrinkage and autogenous shrinkage (Figure 5c), it can be concluded that thermal shrinkage is the main cause of early transverse cracks. Therefore, in a high-temperature environment, the shrinkage phenomenon occurs in the roadway wall concrete under the influence of temperature, resulting in cracks in the shotcrete.

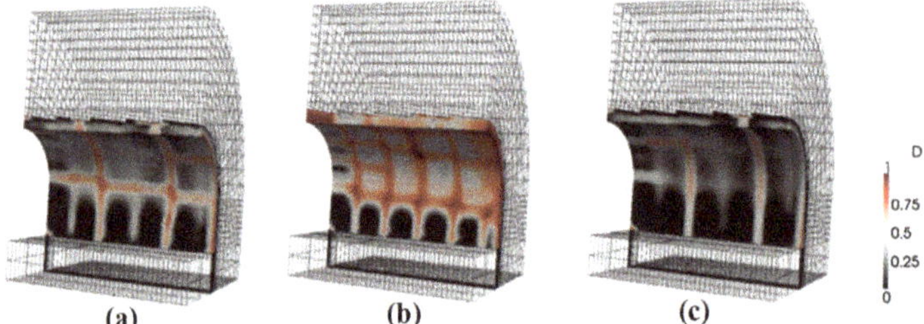

Figure 5. Damage field caused by different phenomena after 360 h (heat shrinkage and self-contraction) [49]. (**a**) Creep is considered; (**b**) creep is ignored; and (**c**) only thermal shrinkage is considered.

3. Mechanical Properties and Microstructure of Shotcrete under High Temperature

Shotcrete technology is the preferred technology for roadway support at present. Under high temperature, however, the mechanical properties of shotcrete will be changed. This section discusses the performance of shotcrete under different factors under high temperature.

3.1. Variation Law of Compressive and Flexural Strength of Shotcrete

With the increasing number of high-temperature construction sites, some scholars have studied the influence of formation temperature on shotcrete, firstly through the change of compressive and flexural strength. As shown in Figure 6, Liu et al. [50] found that the compressive strength and splitting tensile strength of shotcrete increased with the increase of curing age through standard curing in a 40–100 °C dry and hot environment. Under a dry and hot environment of 60–100 °C, it shows an increasing trend before curing for 7 d, and has an obvious downward trend after 7 d. After curing for 28 d, the compressive strength under high temperature is lower than that under standard curing. At 7 d and 28 d, the splitting tensile strength of concrete decreases with the increase of curing temperature. Similarly, Zhu et al. [51] found that the initial setting time and final setting time of high-strength shotcrete shortened with the increase of curing temperature, and analyzed the compressive strength of high-strength shotcrete with the curing ages of 3 d and 28 d. The results showed that the overall compressive strength presents an increasing trend, but when the temperature is too high, the compressive strength of shotcrete has an obvious shrinkage phenomenon.

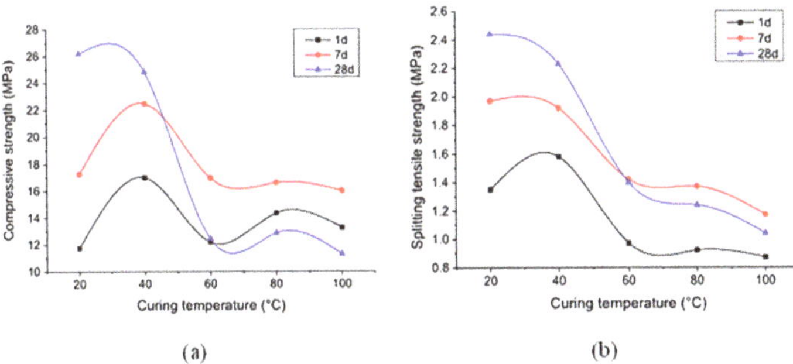

Figure 6. Mechanical properties of specimens under various working conditions: (**a**) compressive strength test, (**b**) splitting tensile strength test [50].

The above research only expounds the deterioration law of shotcrete as a macro phenomenon, but does not explain the causes of deterioration from the micro level.

In order to deeply study the influence of high temperature on the deterioration of shotcrete, scholars analyzed the deterioration mechanism of shotcrete from the perspectives of carbonation depth and early hydration products of cement.

In view of the critical size effect of concrete under high temperature, scholars have proposed using acoustic emission technology to evaluate the internal structural defects and damage of concrete [52–54]. Relevant conclusions show that there is a negative correlation between temperature and acoustic emission signal. Before the compressive strength of concrete has reached the critical point, the acoustic emission number and energy of concrete showed an increasing trend, until the concrete members were completely destroyed [55].

Li et al. [56] studied the carbonation degree of shotcrete under high temperature, and noted that the compressive strength of shotcrete decreased with the increase of temperature. Results showed that the carbonation depth of shotcrete at a high temperature of 50 °C was more than 4 times higher than that at a normal temperature, which easily reduced the alkalinity and damaged the sealing structure of the shotcrete. Xie et al. [57] found serious cracks on the surface of 100 °C high-temperature shotcrete, and preliminarily explained the phenomenon of shrinkage of shotcrete compressive strength. The early hydration products formed too fast, resulting in disordered accumulation of products, which made the distribution uniformity of the shotcrete's internal structure worse. The later high temperature caused the hydration products to move strongly, and most of the hydration products deposited near the larger aggregate. Figure 7a shows the X-ray diffraction (XRD) patterns of shotcrete cured at 80 °C for 3 h, 4 h, 5 h, and standard curing. The AFt diffraction peak decreases and the C-S-H and C₃S diffraction peaks increase with the curing time at high temperature. This showed that short-term high-temperature curing could strengthen the compactness of shotcrete. Meanwhile, Wang et al. [58] also detected C-S-H phase at 0–60 °C, showing an increasing trend. In addition, many scholars have explained that the increase of shotcrete compactness was due to the gradual loss of shotcrete crystal water in the gradually increasing temperature gradient, and a large amount of ettringite (Aft) would be transformed into monosulfide calcium sulphoaluminate (Afm) [53,59].

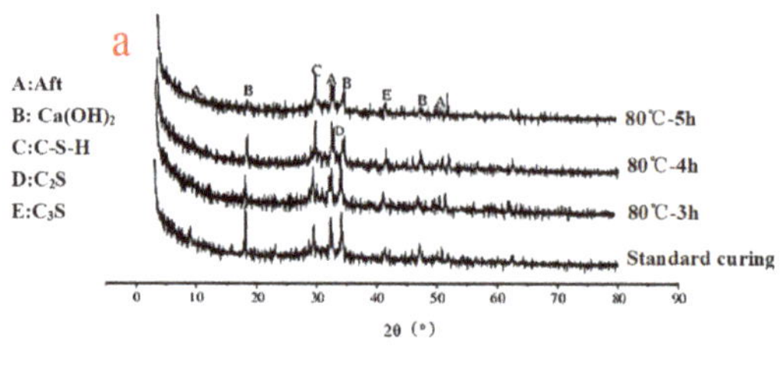

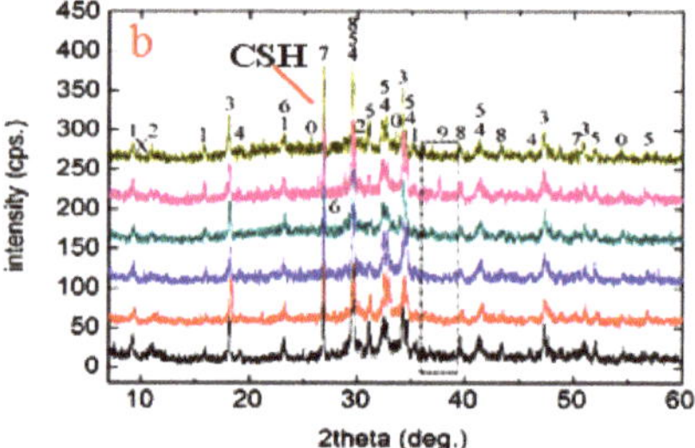

Figure 7. XRD spectra of different curing systems: (**a**) different curing time (**b**) Different curing temperatures [57,58].

Calvo et al. [60] carried out microscopic analysis of shotcrete at 90 °C temperature, and found that the rate of C/S in C-S-H gel increased slightly after thermal curing, while the ratio of (A + F)/C and Sulphates/C decreased, indicating that the properties of C-S-H gel changed under high temperature (mainly sulfate content), as shown in Table 1.

Table 1. Chemical composition of C-S-H gel formed in concrete under different conditions [60].

Curing	Concrete Type	C–S–H gels Composition		
		C/S	A + F/C	Sulphates/C
Standard	H	1.68 ± 0.06	0.145 ± 0.033	0.104 ± 0.048
	F	1.84 ± 0.29	0.159 ± 0.057	0.061 ± 0.021
Heat curing	H	1.84 ± 0.12	0.124 ± 0.026	0.074 ± 0.011
	F	2.04 ± 0.35	0.093 ± 0.004	0.057 ± 0.015

H: Cement content 100%; F: 20% limestone instead of 20% cement.

A group of scholars have studied the performance of shotcrete under ultra-high temperatures (the temperatures were higher than 150 °C), and found that the compressive strength and flexural strength of shotcrete continue to decline with the increase of temperature, and that the damage to shotcrete's internal structure caused by ultra-high temperatures is very serious [61–63]. Under ultra-high temperatures, calcium hydroxide and wollastonite in shotcrete decompose to generate a large amount of calcium oxide, which leads to the instability of the shotcrete's internal structure. Wang et al. [64] studied the influence of temperature and humidity on shotcrete, and found that low-temperature and high-humidity curing conditions were conducive to the development of shotcrete strength at later stages.

The above research lacks studies on the micro void structure of shotcrete. As we all know, the change of micro void structure affects the mechanical properties of shotcrete. With the development of new detection technology, scholars use SEM, CT, and the mercury intrusion method (MIP) to measure the change of voids in shotcrete, so as to further explain the influence mechanism of high temperature on concrete deterioration.

Furthermore, scholars have studied the microstructure of shotcrete under high temperature. Figure 8 show SEM images of ceramsite shotcrete at different temperatures. At 20 °C, the shotcrete presents a particle aggregation state, and the structure is stable. At 40 °C, the shotcrete structure is gradually destroyed, and the particles are granular in structure. At a temperature of 60 °C, the concrete particles are small, there are a lot of micro pores, the structure is loose, and the degree of hydration is low, which leads to a decrease in shotcrete strength. The reason is that ceramsite is a microporous medium, and with the increase of temperature, it accelerates the water evaporation of shotcrete, so the hydration structure of the shotcrete is relatively dispersed.

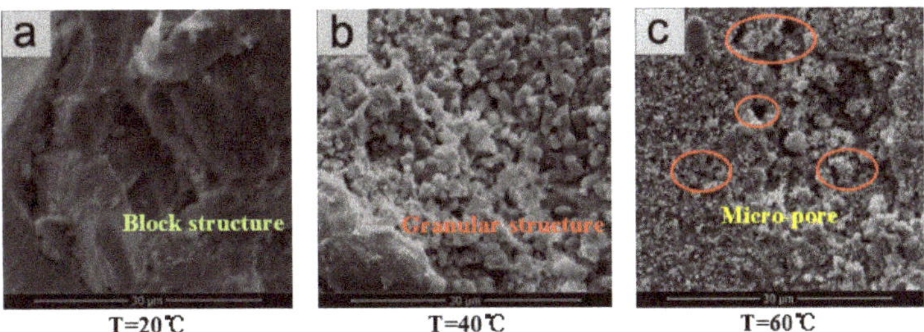

Figure 8. SEM photos of ceramsite concrete curing at different temperatures [65].

Chen and Wei et al. [66,67] found that there was a correlation between the fractal dimension of pore volume and the strength by establishing the fractal model; that is, the more pores there are and the larger they are, the lower the strength of the shotcrete. Some studies have also shown that the permeability of shotcrete could be improved by increasing the integral dimensions of the pores [68]. İlhami et al. [69] studied the performance of shotcrete under high temperature, and concluded that there was an obvious correlation between ultrasonic pulse velocity and compressive strength, that is, the use of ultrasonic pulse velocity method was a feasible method to estimate the high-temperature compressive strength of shotcrete cover. Yang et al. [70] explained that in a hot and humid environment, the expansion stress in AFt caused by too high of a temperature led to structural damage and weakening of the mechanical properties of shotcrete. Shen et al. [71] used the mercury intrusion method (MIP) to study the change of shotcrete porosity at different temperatures, and concluded that the porosity of shotcrete had a strong correlation with the compressive strength; that is, the higher the porosity of shotcrete, the lower the compressive strength. This was expressed by Schiller [72]:

$$\sigma = \text{Dln}\left(\frac{\sigma_0}{P}\right) \tag{3}$$

where σ is the compressive strength of concrete, P is the porosity, σ_0 is the ideal compressive strength when the porosity is 0, and D is the empirical constant.

At present, in terms of the mechanical properties of shotcrete, scholars have studied the impact of low to ultra-high temperatures on concrete. The span of temperature basically covers the range of environmental changes on site. Scholars have not only studied the macro cracks, but also deeply analyzed the deterioration mechanism of shotcrete from the perspective of the micro void structure. However, the correlation between the macro phenomenon and the microstructure of shotcrete in a high-temperature environment is little understood. It is necessary to further strengthen the relationship between the two, especially the macro and micro relationship model, so as to better provide a theoretical basis for research on shotcrete in high-temperature environments.

3.2. Bond Strength between Shotcrete and Coal (Rock) Interface

The bond between shotcrete and coal (rock) interface has been a key problem in roadway support. It was necessary to further study the mechanical properties of the interface under high temperature [73,74]. As shown in Figure 9, Ma et al. [75] studied the relationship between surrounding rock and shotcrete cohesion based on a high-ground-temperature simulation experiment. By setting the temperature range at 50–90 °C, the bond strength decreases with the increase of rock wall temperature. At the same time, when the temperature exceeds the critical value, the bond performance of shotcrete will shrink. In addition, CT scanning was performed on the surface of shotcrete at 50 °C and 90 °C. It was observed that the surface deterioration of shotcrete at 90 °C was more serious than that at 50 °C, with a large number of pores, as shown in Figure 10.

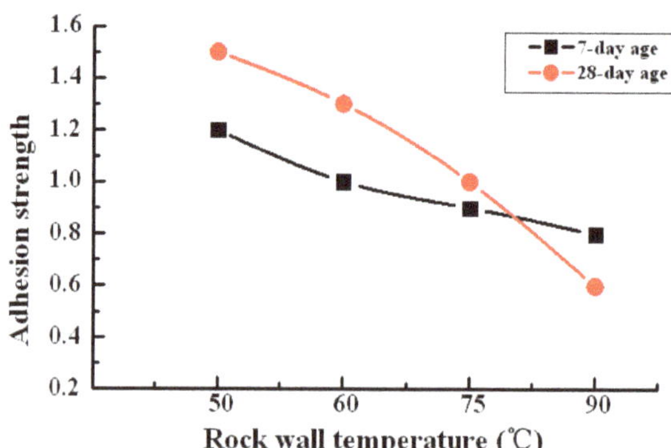

Figure 9. Data of bond strength between surrounding rock and shotcrete under different working conditions [75].

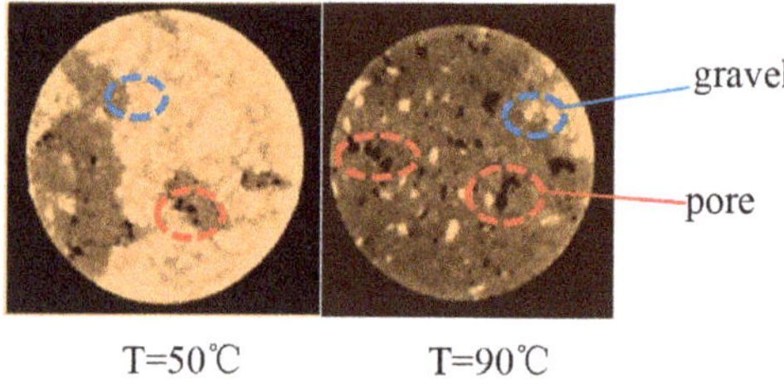

Figure 10. CT scan image of bonding surface [75].

Yang et al. [16] studied the bond strength of shotcrete under 70 °C working conditions in a hot and humid environment, and concluded that a dry and hot environment had the greatest impact on the performance of shotcrete, and would accelerate the retrogradation of shotcrete [19,76]. Su et al. [77–79] carried out finite element analysis on the temperature field of shotcrete rock slab, and found that temperature determined the damage degree of bond strength between shotcrete and rock slab. The heat and the moisture in shotcrete were mainly lost and evaporated through the side wall of the specimen. As shown in Figure 11, the internal pores of shotcrete under various temperature conditions was analyzed. It was determined that the pore area is proportional to the development of the temperature. For one thing, the high temperature led to a rapid evaporation of water in the shotcrete, and pores formed during the hydration process of the cement; for another, the impact force generated in the process of shotcrete spraying caused the shotcrete to be mixed with air, thus producing bubbles, resulting in an increase in the number of pores in shotcrete under high temperature.

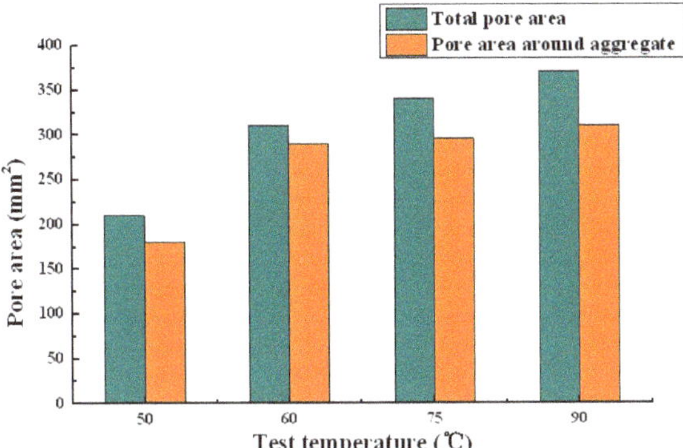

Figure 11. Comparison of total pore and surrounding pore of aggregate at each temperature [79].

Fan et al. [80] studied the bond strength of shotcrete under high temperature by the. splitting method and drawing method, as shown in Figure 12. Because the splitting failure was accompanied by shear phenomenon, the bond strength of shotcrete measured by splitting method was better than that by drawing method under high temperature. It was also found that the cracks on the cemented surface of concrete due to autogenous shrinkage at a high temperature of 60 °C and thermal decomposition still occurred in the transition zone between shotcrete and rock. Duan et al. [81] pointed out in the paper that relevant scholars believe that 75 °C is the critical temperature for thermal damage of shotcrete, and beyond this temperature, shotcrete will lose its adhesion.

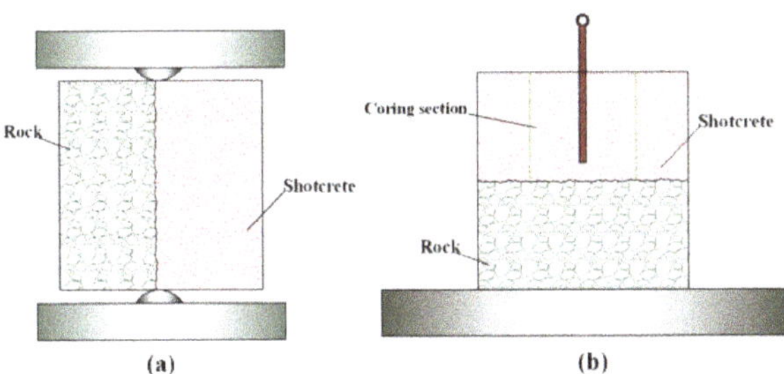

Figure 12. Schematic diagram of bond strength test: (a) splitting method and (b) core drilling and drawing method.

Tang et al. [82] carried out three-dimensional CT scanning on shotcrete to observe the agglomeration phenomenon of shotcrete under different temperatures. It was found that the pore and micro crack zone of shotcrete block at 90 °C was much higher than that at 50 °C. Results showed that the high temperature accelerated the hydration reaction of cement and the disorderly overlapping between molecules, which led to the intensification of autogenous shrinkage of shotcrete and the poor adhesion of hydration products generated, as shown in Figure 13. In addition, the local stress in the shotcrete would produce new cracks again, which would weaken the bond performance of the internal structure of shotcrete.

Scholars have studied the effect of void structure on bond performance of shotcrete by using CT technology, which is of certain significance for analyzing the mechanical failure of shotcrete. However, there are relatively few studies on the interface between shotcrete and rock (coal) based on macro and micro analysis, and in particular, the effect of parameters of different sprayed surfaces (such as roughness, density, and chemical composition) on the bonding performance of shotcrete needs to be further studied.

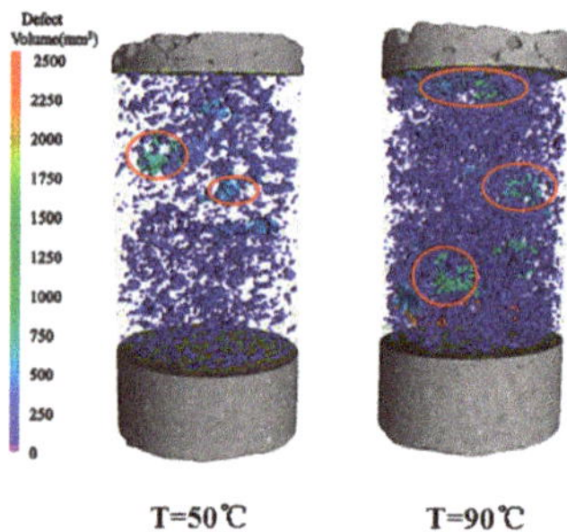

Figure 13. Porosity distribution in shotcrete based on CT image [82].

3.3. Other Mechanical Properties of Shotcrete

There are many failure forms of shotcrete and roadway walls in roadways [83,84]. The shear failure model of shotcrete and rock wall under a standard and high-temperature environment is shown in Figure 14. The internal structure of shotcrete was seriously damaged under high temperature, and the damaged particles presented an irregular state. However, the structure of shotcrete was stable after shear failure in a standard environment. Some scholars have studied the performance of shotcrete under shear failure and other forms. For example, Tang et al. [85] studied the influencing factors of shear strength at the cementation surface between shotcrete and granite under conditions of high and variable temperatures. It was concluded that normal stress has the greatest influence on the shear strength, followed by temperature and humidity, and the surface roughness of granite had the least influence. Moreover, the shear strength increased first, and then decreased with the curing temperature.

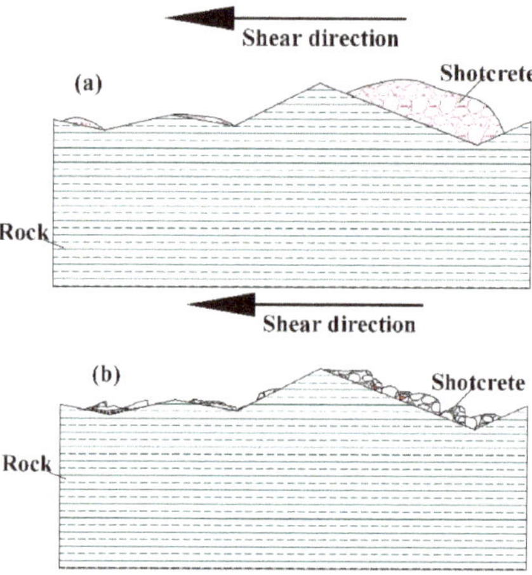

Figure 14. Shear failure model: (**a**) standard environment and (**b**) high temperature environment.

Tong et al. [86] studied the influence of different normal stresses on the interfacial shear strength of shotcrete under high-temperature conditions, and found that the shear strength first increased and then decreased with the increase of temperature, and increased with the increase of normal stress. Furthermore, the maximum shear strengths corresponding to temperature was 60 °C and 80 °C when the relative humidity (RH) were 55% and 95%. Consequently, the results showed that the higher the relative humidity was, the higher the peak shear strength and the peak cohesion were, as shown in Figure 15. Mohamed et al. [87]

studied the limit stress of saturated shotcrete and dry shotcrete under high temperature. It was found that the thermal effect of shotcrete in a saturated state failed due to the gap pressure generated by free water evaporation during the heating stage, which destroyed the shotcrete matrix and led to serious brittle failure of shotcrete.

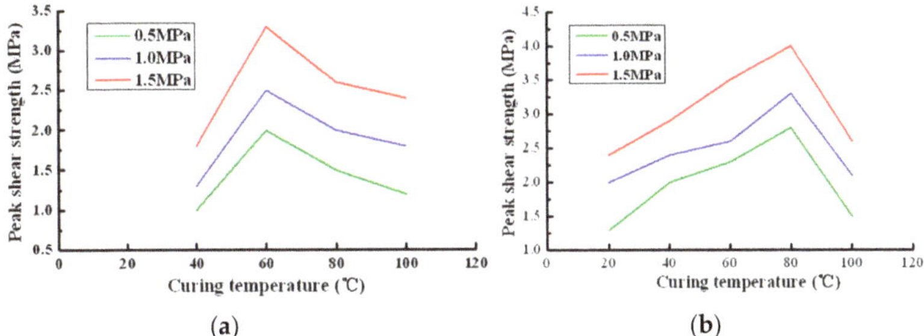

Figure 15. Variation trend of peak shear stress of C30 concrete with temperature: (**a**) 55% RH and (**b**) 95% RH [86].

Pan et al. [88] studied the residual strength of shotcrete under natural cooling and immersion cooling in an ultra-high temperature environment, as shown in Figure 16. The residual strength of shotcrete was higher when the temperature was lower. However, when the temperature continued to rise, the variation law of residual strength was basically the same under the two curing methods. This shows that the cement hydrate decomposes in the shotcrete under ultra-high temperature, and the bond between aggregate and cement stone becomes worse; the temperature inside and outside of the shotcrete was uneven due to water cooling under high temperature, which led to a shrinkage of the internal structure of the shotcrete. It was further explained that the performance change of shotcrete had nothing to do with the curing method when the temperature exceeded 800 °C.

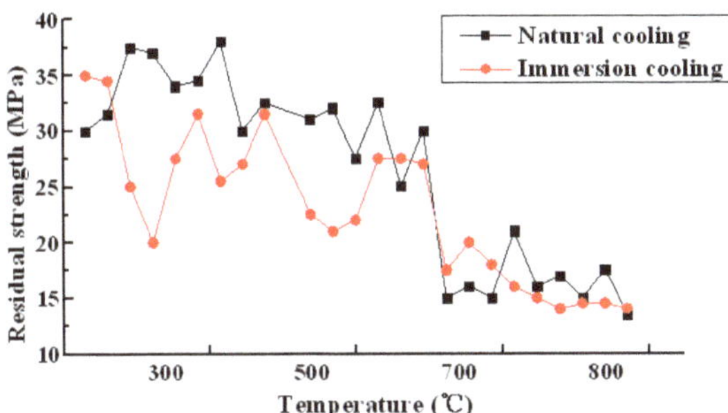

Figure 16. Residual strength of concrete under different cooling modes in a high-temperature environment [88].

There are many studies on the mechanical properties of shotcrete under high temperature, and a large number of research results have been obtained to explain the macro phenomenon from the micro point of view. However, most studies only focus on the single mechanical properties of shotcrete. The overall studies on comprehensive properties are relatively few, and the correlation between various mechanical properties needs to be further discussed. It is suggested that when studying various mechanical properties of shotcrete at high temperature, detailed micro research can be conducted separately to focus on the macro changes of various mechanical properties. This may be a relatively clear research method, rather than performing a microstructure study every time a mechanical property is studied.

4. Methods to Improve the Performance of Shotcrete under High Temperature

4.1. Cooling Technology and Method

At present, refrigeration technology at home and abroad mainly includes air conditioning technology, air cooling technology, ice cooling technology, and thermoelectric glycol technology [89–91], as shown in Figure 17. However, the cost of traditional technology is high, which is undesirable for a construction environment; therefore, it is necessary to solve the problem of high temperature heat damage from the perspective of new technology and new methods. For example, Jin et al. [92] studied the impermeability of concrete under high temperature with a spray-cooling method. It was found that spray-cooling could improve the ambient temperature, but the damage to the surface of the shotcrete resulted in weakening of the mechanical properties of the shotcrete. Luo et al. [93] introduced a solution using dehumidification and cooling technology, in which NaCl solution and $CaCl_2$ solution are selected, and the air is cooled by the differential pressure between solution and air. Some scholars have also studied liquid CO_2 cooling refrigeration technology for ventilation, and found that the air flow temperature in the air duct gradually decreased [94], and the cooling effect was very good. Reducing the high temperature in a construction environment and increasing the air humidity in the construction site can be done by physical or chemical means. A humid environment is conducive to the recovery of high-temperature-damaged concrete strength [95]. In addition, because of the huge geothermal productivity in a high-temperature environment, geothermal refrigeration technology could also be used to improve the construction environment. Table 2 summarizes some refrigeration technologies and compares their advantages and disadvantages.

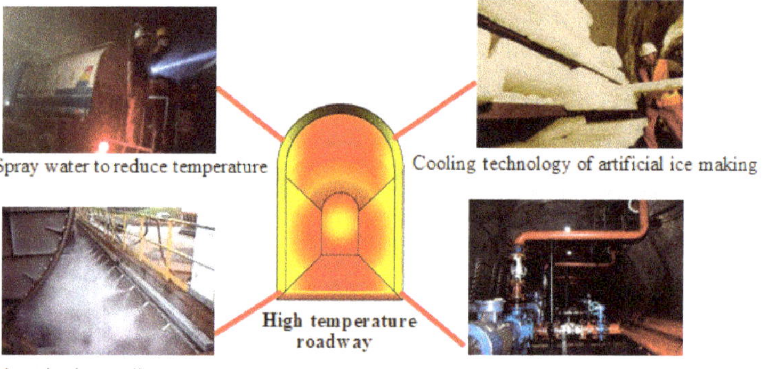

Figure 17. Refrigeration technology in high temperature construction environments.

4.2. Experimental Study on Performance Optimization of Shotcrete

From the safety point of view, in construction in a deep environment, we should not only ensure the stability of the roadway structure, but also improve the site construction environment. Therefore, some scholars have carried out optimization research on shotcrete. He et al. [96] studied thermal insulation shotcrete, and found that the heat insulation mechanism was to increase the temperature difference between the rock wall and construction roadway by using lightweight aggregate with osteoporosis and high porosity as the raw material. Moreover, the test showed that the shotcrete had a certain cooling effect on the construction roadway. Mohd and Zhou et al. [97,98] found that shotcrete still had strong mechanical properties under ultra-high temperature by adding fly ash and metakaolin into shotcrete. Yang et al. [99] found that 67% of the residual strength could be maintained at 1000 °C when steel fiber was added into high-performance shotcrete, which could effectively inhibit the high-temperature cracking of shotcrete.

Table 2. Comparison of advantages and disadvantages of refrigeration technology.

Refrigeration Technology		Advantage	Disadvantage
Air conditioning refrigeration technology	Air cooling technology	Using air as a refrigeration medium, the system produced no pollution in the environment, and the system structure was simple	Large refrigeration equipment, inconvenient to move and install, poor refrigeration capacity, and high refrigeration cost
	Water refrigeration technology	The refrigeration range was wide, and the cooling effect was obvious	In the process of cooling, there will be a temperature jump
	Geothermal refrigeration technology	Making full use of waste heat energy, environmental protection and energy saving, low operating cost	The refrigeration range was limited, so it was necessary to establish multiple refrigeration systems
	Ice refrigeration technology	High cold storage capacity and high heat exchange efficiency	The conveying process was easily blocked, and the melting speed was low
Non-mechanical cooling technology	Spray-cooling technology	It had the advantages of fast cooling rate, low cost, and good dust removal effect	It was suitable for a tunneling working face with small working range, few operators, small cooling capacity, and serious heat damage
	Liquid CO_2 refrigeration technology	The conveying distance was more than 1000 m, the cooling effect was good, and the cooling system was flexible	When the air volume of the working face is large, the cooling effect will be affected
	Solution dehumidification and cooling	It had high dehumidification efficiency, controllable dehumidification capacity, and low operation cost. It could remove dust, bacteria, and other harmful substances in the air	The initial investment cost of the solution dehumidification unit was high

However, some researchers also found that the mechanical properties of thermal insulation shotcrete mixed with vitrified microbubbles had decreased, which was explained by the formation of a large number of porous structures when vitrified beads were added into shotcrete (Figure 18). In addition, there were also scholars who added inorganic materials such as expanded perlite and silica fume into shotcrete to study its thermal insulation performance; the inorganic materials are shown in Figure 19. For example, Liu et al. [100] studied thermal insulation shotcrete by replacing sand with expanded perlite, and found that dry-spraying technology had a better thermal insulation effect. Pang et al. [101] developed a new type of shotcrete thermal insulation material by adding ceramsite, vitrified microsphere, and fly ash into shotcrete. According to the test, the thermal conductivity was between 0.1837 and 0.2533 w/(m·K), which had certain heat insulation performance and could improve the high-temperature environment. Some scholars have put forward the idea of spraying polystyrene foam for shotcrete maintenance, reducing the temperature difference between the inside and outside of shotcrete, so as to prevent shotcrete cracks [102]. Lei et al. [103] studied the performance change of shotcrete under cement temperatures. The mechanical properties of shotcrete could be improved, and the early setting time could be shortened, by heating the cement at a certain temperature before the shotcrete was prepared. Jiang et al. [104] added plant fiber into shotcrete and found that the thermal conductivity of shotcrete decreased by 20.61% compared with that of ordinary shotcrete, and the fiber inhibited the generation of microcracks to a certain extent [105].

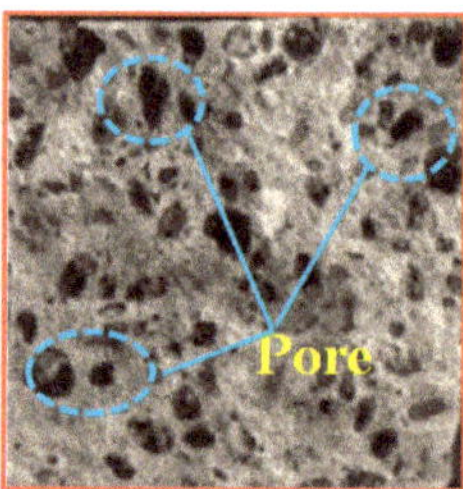

Figure 18. Cross section of thermal insulation concrete material [106].

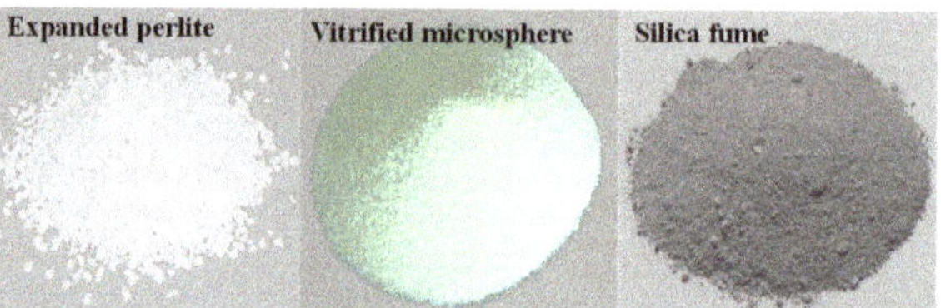

Figure 19. Inorganic materials for performance optimization of shotcrete.

Cui et al. [107,108] studied the performance changes of shotcrete mixed with hook-end steel fiber (SF-HE), wave-shaped steel fiber (SF-W), and basalt fiber (BF) at 100 °C, and found that the total porosity and harmful porosity of shotcrete decreased significantly after adding fiber, and the optimization effect of steel fiber on pore structure was better than that of basalt fiber. However, the porosity of shotcrete increased after the steel fiber and silica fume were mixed, which indicated that the pozzolanic activity of silica fume fails under high temperature, as shown in Figure 20. Chu et al. [109] found that the toughness of foam-fiber-reinforced shotcrete decreased, and the thermal conductivity decreased. This was because foam filled in the void of shotcrete and directly cut off the heat transfer path. Patrick et al. [110] studied the structural performance of shotcrete and concluded that the shotcrete containing alkaline accelerator had low thermal diffusivity and better thermal insulation performance. Some scholars found that the degree of C-S-H gel polymerization and chain length increased at the cemented surface when the shotcrete was mixed with silica fume. A large number of CH crystals generated C-S-H gel, which filled in the interface pores, which enhanced the interfacial strength [111,112]. This could be expressed as $Ca(OH)_2 + SiO_2 + H_2O \rightarrow C\text{-}S\text{-}H$. Benarchid et al. [113] analyzed the vulcanization properties of waste rock and concluded that it had good safety. Francesco et al. [114] proposed that the performance of shotcrete could be optimized by changing the density of the material to reduce the pore pressure and the gravitational potential energy of shotcrete.

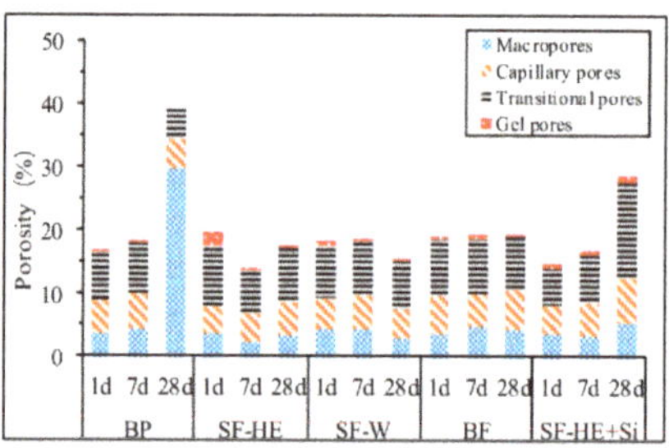

Figure 20. Porosity of different pores under different conditions [108]. BP: non-fiber material; SF-HE: 1% hook-end steel fiber; SF-W: 1% corrugated steel fiber; BF: 0.1% basalt fiber; SF-HE+Si: 1% steel fiber and 5% silica fume.

5. Challenges

At present, despite the plentiful results obtained in the field of high-temperature shotcrete research, most of the research is still in the field of tunnel engineering, and there are few studies in other underground engineering fields (such as mine engineering). Moreover, the environments of mine engineering and tunnel engineering are not the same. The mining depth of mine engineering is far greater than that of tunnel engineering. The mining environment is more severe, and the high-temperature environment is more complicated. Therefore, the research on sprayed concrete in the field of underground engineering still faces great challenges, as shown in Table 3.

Table 3. Challenges of shotcrete performance under high temperature.

Challenges	Current Technologies	Future Challenges
High temperature detection technology research	Thermocouple temperature measurement, infrared temperature measurement, and other technologies. However, the temperature can change in the hot and humid environment of the roadway, and a complete test system has not been formed.	Develop a new temperature measurement method, establish a temperature measurement system, predict and forecast the high temperature area of the roadway, ensure that the construction environment is in a balanced state, and reinforce the area where the roadway sprayed concrete may be damaged in advance.
Experimental device for performance of shotcrete	Shotcrete is made by pouring or spraying. Then, constant temperature and a humidity curing box or high-temperature drying box are used to simulate high-temperature environment curing. There will be a gap in the simulation of the high-temperature environment, which will affect the experimental results.	Research and development of shotcreting equipment and performance testing system in high temperature. Accurately simulate high-temperature environment to improve the authenticity of research results. Provide reliable results for improving the development of infrastructure under high temperature.
Shotcrete spray layer structure	The existing research is based on the study of the adhesion of shotcrete under the action of temperature, but it has not involved research on the shotcrete spraying layer in the process of roadway reinforcement under high temperature.	Study the performance change trend of high-temperature roadway shotcrete with the thickness of shotcrete layer. Get the optimal spray layer thickness under high temperature. Study the adhesion between spray layers. Research and develop high-efficiency, environmentally friendly, and low-cost adhesives
Performance of shotcrete in special environments	Existing research is based on the permeability of shotcrete, and there is basically no relevant literature on the high-temperature conditions of the construction roadway in a high-water-spray environment.	Study the mechanical properties of shotcrete in high-temperature acid and alkaline environments. Investigate the non-linear relationship between the compatibility of the sprayed concrete and the cemented surface of the high-water-spray area and the interface's mechanical properties.
Microscopic properties of shotcrete	The existing research has explained the deterioration performance of shotcrete at high temperature from the micro level, but it is still not perfect. There is no detailed study on the crack development process and deterioration mechanism of concrete in a high temperature environment.	Study the microscopic changes of shotcrete under high temperature. Establish a damage model of shotcrete under high temperature. Study the impact properties of shotcrete layers under high temperature. Establish a damage prediction model for shotcrete layer structure under high temperature, using acoustic emission technology to predict the development trend of shotcrete deterioration in advance.

Table 3. *Cont.*

Challenges	Current Technologies	Future Challenges
Optimization of shotcrete	Existing research is mainly based on studies of thermal insulation sprayed concrete, which is mainly improved from the material ratio by adding inorganic materials. Although it has a certain thermal insulation effect, its thermal insulation performance is insignificant for the ultra-high temperature construction environment.	Study the modification of shotcrete. Choose low-quality, low-thermal-conductivity materials to reduce the potential energy of concrete. Preliminary research on the high temperature resistance between silica aerogel and shotcrete.

6. Conclusions

In recent years, a large number of scholars have carried out various studies on the mechanical properties and micro characteristics of shotcrete at high temperature. This paper systematically discusses the evolution mode and process of concrete performance under high temperature:

(1) This paper first reviewed the causes of the formation of high temperature environments, and pointed out that formation temperature was a kind of heat conduction mode. The leading role of rock thermal conductivity on temperature transmission was determined by introducing the multi-dimensional morphological formula in the process of heat conduction.

(2) The mechanical properties and micromechanical properties of shotcrete under high temperature were reviewed. Results concluded that the mechanical properties (including compressive strength, tensile strength, bond strength, shear strength) of shotcrete were affected by the critical temperature: before the critical temperature, the mechanical properties of shotcrete showed an increasing trend with the increase of temperature; after the critical temperature, the mechanical properties of shotcrete appeared to show the phenomenon of shrinkage. Through microscopic analysis, multiple studies have shown that when the temperature exceeds the critical temperature, the internal molecules of the shotcrete move violently and the molecules overlap in a disorderly way, the hydration reaction of shotcrete cement is terminated prematurely, and the brittle de-formation is enhanced, which leads to the weakening of shotcrete strength.

(3) The cooling technology and performance optimization of shotcrete at high temperature are summarized. It is concluded that taking cooling measures for a high-temperature construction environment will increase the recoverability of concrete after deterioration and reduce the deterioration degree of shotcrete. In terms of optimizing the performance of shotcrete, adding inorganic materials, such as vitrified microbeads, foam fibers, expanded perlite, and silica fume, will improve the heat insulation and heat resistance of shotcrete.

(4) The research status of shotcrete technology under high temperature was analyzed from the aspects of equipment, materials, and technology. It was found that the current temperature measurement system, high-temperature simulation equipment, and material ratio had limitations. At the same time, the challenges that high-temperature shotcrete faces in terms of the process structure, performance optimization, and its application in special engineering fields were summarized.

Funding: This study was funded by the National Key Research and Development Plan of the 13th Five-Year Period (Grant No. 2017YFC0805203); National Natural Science Foundation of China (Grant No. 51974177, 51934004); and Natural Science Foundation of Shandong (Grant No. ZR201801280006, ZR2019QEE007, ZR2019MEE115).

Institutional Review Board Statement: Not applicable.

Informed Consent Statement: Not applicable.

Data Availability Statement: Not applicable.

Conflicts of Interest: The authors declare that they have no known competing financial interest or personal relationship that could have appeared to influence the work reported in this paper.

References

1. Xiamen Academy of Building Sciences Group Co., Ltd. *Technical Specification for Application of Sprayed Concrete JGJ/T 372*; China Construction Industry Press: Beijing, China, 2016.
2. Burns, B.D. *Characterization of Wet-Mix Shotcrete for Small Line Pumping*; Laval University: Québec, QC, Canada, 2008.
3. Brooks, J.W. Shotcrete and its application. In *Shotcreting Proceedings*; South African Institute of Mining and Metallurgy School: Johannesburg, South Africa, 1998.
4. Chen, L.; Li, P.; Liu, G.; Cheng, W.; Liu, Z. Development of cement dust suppression technology during shotcrete in mine of China-A review. *J. Loss Prev. Process. Ind.* **2018**, *55*, 232–242. [CrossRef]
5. Liu, G.; Chen, L.; Cheng, W.; Huang, Y. Research on pump primers for friction reduction of wet-mix shotcrete based on pre-creating lubricating layer. *Adv. Mater. Sci. Eng.* **2017**, *2017*, 1–12.
6. Zhu, K.; Zhou, Y.; Wang, P.; Liu, J.; Shi, C. Study on monitoring technique and distribution regularity of geothermal field in deep mining. *Nonferrous Met.* **2016**, *68*, 90–94.
7. Chen, X. Prediction and analysis of ground temperature in deep mining of metal mine. *Sci. Technol. Inf.* **2017**, *15*, 94–111.
8. Zhang, C.; Zhao, L.; Wu, Z. Depth characteristics of Curie surface in the Eastern seas of China. *J. Shandong Univ. Sci. Technol.* **2019**, *38*, 10–17.
9. Wang, X.; Li, X.; Kong, F.M. Metamorphism and tectonic evolution of the sedimentary-matrix mélange in the central Yarlung Zangbo Suture Zone, Southern Tibet. *J. Shandong Univ. Sci. Technol.* **2020**, *39*, 16–27.
10. Yao, H.; Xu, C.; Wu, H.; Zheng, Z.; Sun, X.; Wang, H.; Chai, J.; Li, Y. Study on geothermal parameters of mid-deep and deep mines in Inner Mongolia and Shaanxi mining area. *China Coal* **2017**, *43*, 91–96.
11. Wang, C. Analysis of thermal damage factors in deep mine. *Zhongzhou Coal.* **2013**, *40*, 42–105.
12. Guo, Z. Current situation and technical problems of deep mining in China's coal mines. *Coal* **2017**, *26*, 58.
13. He, M.; Guo, P. Deep rock mass thermodynamic effect and temperature control measuresd. *Chin. J. Rock Mech. Eng.* **2013**, *32*, 2377–2393.
14. Shengai, C.; Pin, L.; Jiao, S.; Enqi, C.; Chen, G.; Bing, Z. Experimental study on mechanical and microstructural properties of cement-based paste for shotcrete use in high-temperature geothermal environment. *Constr. Build. Mater.* **2018**, *174*, 603–612.
15. Lee, C.-H.; Wang, T.-T.; Chen, H.-J. Experimental study of shotcrete and concrete strength development in a hot spring environment. *Tunn. Undergr. Space Technol.* **2013**, *38*, 390–397. [CrossRef]
16. Yang, H. *Performance Research and Structural Behavior Analysis of Shotcrete in Thermal Damage Tunnel*; Southwest Jiaotong University: Chengdu, China, 2013.
17. Wang, J.; Niu, D.; Wang, Y.; He, H.; Liang, X. Chloride diffusion of shotcrete lining structure subjected to nitric acid, salt–frost degradation, and bending stress in marine environment. *Cem. Concr. Compos.* **2019**, *104*, 103396. [CrossRef]
18. Wang, J.; Niu, D.; Ding, S.; Mi, Z.; Luo, D. Microstructure, permeability and mechanical properties of accelerated shotcrete at different curing age. *Constr. Build. Mater.* **2015**, *78*, 203–216. [CrossRef]
19. Cui, S.A.; Li, J.W.; Ye, Y.Z.; Yang, H.Y. Bond Strength of shotcrete with rock in dry and hot environment of high ground temperature tunnel. *J. Build. Mater.* **2013**, *4*, 663–666.
20. Dong, W.; Yang, D.; Zhou, X.; Kastiukas, G.; Zhang, B. Rimental and numerical investigations on fracture process zone of rock–concrete interface. *Fatigue Fract. Eng. Mater. Struct.* **2017**, *40*, 820–835. [CrossRef]
21. Yang, C.; Chen, J. Fully noncontact nonlinear ultrasonic characterization of thermal damage in concrete and correlation with microscopic evidence of material cracking. *Cem. Concr. Res.* **2019**, *123*, 105797. [CrossRef]
22. Lu, T.; Zhao, G.; Lin, Z. Experimental study on mechanical properties of long standing concrete after exposure to high temperature. *J. Build. Struct.* **2004**, *25*, 63–70.
23. Lu, T.; Zhao, G.; Lin, Z.; Yue, Q.R. Microscopic analysis of long standing concrete after high temperature. *J. Build. Mater.* **2003**, *6*, 135–141.
24. Kjellsen, K.O. Heat curing and post-heat curing regimes of higher-performance concrete: Inflfluence on microstructure and C–S–H composition. *Cem. Concr. Res.* **1996**, *26*, 295–307. [CrossRef]
25. Akca, A.H.; Zihnioğlu, N.Ö. High performance concrete under elevated temperatures. *Constr. Build. Mater* **2013**, *44*, 317–328. [CrossRef]
26. Zonno, G.; Aguilar, R.; Boroschek, R.; Lourenço, P.B. Analysis of the long and short-term effects of temperature and humidity on the structural properties of adobe buildings using continuous monitoring. *Eng. Struct.* **2019**, *196*, 1–21. [CrossRef]
27. Kang, F.; Li, J.; Zhao, S.; Wang, Y. Structural health monitoring of concrete dams using long-term air temperature for thermal effect simulation. *Eng. Struct.* **2019**, *180*, 642–653. [CrossRef]
28. Liu, L.; Zhang, J. Thermal insulation composite material for governance of underground thermal hazard and its application. *J. Shandong Univ. Sci. Technol. Nat. Sci.* **2017**, *36*, 46–53.
29. Wang, Z.; Yin, Y.; Zhao, T.; Ma, J.; Wu, W. Numerical simulation of the mechanism of the hole expansion and anchorage and support control in soft rock roadway. *J. Shandong Univ. Sci. Technol. Nat. Sci.* **2021**, *40*, 9.

30. Xin, S.; Meng, X.; Qu, Y.; Chen, X. Mechanism analysis of heat harm in large dip angle coal mining working face. *J. Shandong Univ. Sci. Technol. Nat. Sci.* **2016**, *35*, 42–48.
31. Lin, X. Formation and distribution of ground temperature field in mines. *Inn. Mong. Sci. Econ.* **2011**, 62–63.
32. Conti, M.; Pata, V.; Pellicer, M.; Quintanilla, R. On the analyticity of the MGT-viscoelastic plate with heat conduction. *J. Differ. Equ.* **2020**, *269*, 7862–7880. [CrossRef]
33. Zhang, Y.-L.; Li, T.; Mou, R.-T.; Wang, C.-Z.; Yuan, F.-C. Study of underground temperature prediction method for deep exploitation of mines under complicated circumstances. *J. Qingdao Technol. Univ.* **2015**, *36*, 1–6.
34. Zhou, Y.; Gu, L.; Li, L.; Wu, L. Study on thermal conductivity of rock and thermophysical properties of coal. *Energy Technol. Manag.* **2011**, *143*, 134–135.
35. Luo, J.; Qi, Y.; Zhao, Q.; Tan, L.; Xiang, W.; Rohn, J. Investigation of flow and heat transfer characteristics in fractured granite. *Energies* **2018**, *11*, 1228. [CrossRef]
36. Luo, J.; Zhu, Y.; Guo, Q.; Tan, L.; Zhuang, Y.; Liu, M.; Zhang, C.; Xiang, W.; Rohn, J. Experimental investigation of the hydraulic and heat-transfer properties of artificially fractured granite. *Sci. Rep.* **2017**, *7*, 39882. [CrossRef]
37. Zhan, F.L.; Cai, M.F. Rock heat-transfer model of high-temperature stopes and its solving process. In Proceedings of the International Young Scholars' Symposium on Rock Mechanics, Beijing, China, 28 April–2 May 2008.
38. Tan, Y.; Zhang, Q. Numerical analysis of surrounding rock deformation and failure in deep roadway under the condition of thermal-solid coupling. *J. Shandong Univ. Sci. Technol. Nat. Sci.* **2016**, *35*, 29–37.
39. Liu, G.; Guo, X.; Cheng, W.; Chen, L.; Cui, X. Investigating the migration law of aggregates during concrete flowing in pipe. *Constr. Build. Mater.* **2020**, *251*, 119065. [CrossRef]
40. Liu, G.; Cheng, W.; Chen, L.; Pan, G.; Liu, Z. Rheological properties of fresh concrete and its application on shotcrete. *Constr. Build. Mater.* **2020**, *243*, 118180. [CrossRef]
41. Zhu, P.; Yu, S.; Liu, W.; Sun, Y. Review of the research status of high performance concrete anti-tunnel fire spalling. *Concrete* **2018**, 24–29.
42. Li, Y.; Shi, K.; Chen, Z.; Li, Z. Study on the early crack resistance of high performance concrete under the condition of high temperature curing. *Yellow River* **2017**, *39*, 130–132.
43. Huang, G.; Zhang, R.; Zhang, R.; Hua, L.; Guo, H. Effect of high temperature on mechanical properties of early age concrete. *Railw. Eng.* **2019**, *59*, 140–142.
44. Qin, L.; Li, M.; Ding, J.-N. The effect from curing of high temperature and high humidity to durability of high strength concrete. *J. Northeast. China Inst. Electr. Power Eng.* **2016**, *36*, 18–22.
45. Li, X.-H.; Wang, J.; Duan, Y. Research on the effects of high temperature tunnel of sprayed concrete performance. *J. Hebei Univ. Technol.* **2014**, *31*, 17–20.
46. Hiremath, P.N.; Yaragal, S.C. Effect of different curing regimes and durations on early strength development of reactive powder concrete. *Constr. Build. Mater.* **2017**, *154*, 72–87. [CrossRef]
47. Jung, W.; Choi, S.J. Effect of high-temperature curing methods on the compressive strength development of concrete containing high volumes of ground granulated blast-furnace slag. *Adv. Mater. Sci. Eng.* **2017**, *2017*, 1–6. [CrossRef]
48. Mymrin, V.; Aibuldinov, E.K.; Alekseev, K.; Petukhov, V.; Avanci, M.A.; Kholodov, A.; Taskin, A.; Catai, R.E.; Iarozinski, A.N. Efficient road base material from Kazakhstan's natural loam strengthened by ground cooled ferrous slag activated by lime production waste. *J. Clean. Prod.* **2019**, *231*, 1428–1436. [CrossRef]
49. D'aloia, L.; Benboudjema, F.; Briffaut, M. Effect of fibres on early age cracking of concrete tunnel lining. Part II: Numerical simulations. *Tunn. Undergr. Space Technol.* **2016**, *59*, 215–220.
50. Pin, L.; Shengai, C.; Zihao, L.; Xuefeng, X.; Chen, G. Inflfluence of surrounding rock temperature on mechanical property and pore structure of concrete for shotcrete use in a hot-dry environment of high-temperature geothermal tunnel. *Constr. Build. Mater.* **2019**, *207*, 329–337.
51. Zhu, L.-L.; Zhao, Y. Effect of temperature on the properties of high strength sprayed concrete. *Bull. Chin. Ceram. Soc.* **2017**, *36*, 2424–2429.
52. Naus, D.J. *The Effect of Elevated Temperature on Concrete Materials and Structures—A Literature Review*; Oak Ridge National Laboratory: Oak Ridge, TN, USA, 2006.
53. Carpinteri, A.; Lacidogna, G.; Pugno, N. Structural damage diagnosis and life-time assessment by acoustic emission monitoring. *Eng. Fract. Mech.* **2007**, *74*, 273–289. [CrossRef]
54. Grosse, C.; Ožbolt, J.; Richter, R.; Periškić, G. Acoustic emission analysis and thermo-hygromechanical model for concrete exposed to fire. *J. Acoust. Emiss.* **2010**, *28*, 188–203.
55. Geng, J.; Sun, Q.; Zhang, W.; Lü, C. Effect of high temperature on mechanical and acoustic emission properties of calcareous-aggregate concrete. *Appl. Therm. Eng.* **2016**, *106*, 1200–1208. [CrossRef]
56. Li, Z. *Study on Deterioration Information and Performance Guarantee of Shotcrete in Deep Mine*; Shandong University of Science and Technology: Qingdao, China, 2018.
57. Xie, Z.; Li, X. Effect of curing temperature and curing time on compressive strength of fly ash geopolymer concrete. *Concrete* **2014**, 55–58.
58. Wang, C. *Influence of High Temperature Curing on Strength and Durability of High Strength Concrete for C80 Pile*; Southwest Jiaotong University: Chengdu, China, 2015.
59. Frigione, M.; Mairo, N. Concrete mixing at elevated temperature: Workability, strength and durability. *Cem. Wapno Beton* **2015**, *2015*, 115–129.

60. Calvo, J.G.; Alonso, M.C.; Luco, L.F.; Velasco, M.R. Durability performance of sustainable self compacting concretes in precast products due to heat curing. *Constr. Build. Mater.* **2016**, *111*, 379–385. [CrossRef]
61. Qu, H. *Experimental Study and Numerical Simulation Analysis of Concrete Performance After High Temperature*; Jilin University: Changchun, China, 2018.
62. Deng, M.; Cheng, Y.; Weng, S.; Zhang, Y. Compressive properties and micro-structure of high ductility concrete exposed to elevated temperature. *Acta Mater. Compos. Sin.* **2020**, *37*, 985–996.
63. Hassan, A.; Aldhafairi, F.; Abd-EL-Hafez, L.M.; Abouelezz, A.E.Y. Retrofitting of different types of reinforced concrete beams after exposed to elevated temperature. *Eng. Struct.* **2019**, *194*, 420–430. [CrossRef]
64. Wang, M.; Hu, Y.; Wang, Q.; Tian, H.; Liu, D. A study on strength characteristics of concrete under variable temperature curing conditions in ultra-high geothermal tunnels. *Constr. Build. Mater.* **2019**, *229*, 116989. [CrossRef]
65. Fan, L.; Zhang, Z.; Yu, Y.; Li, P.; Cosgrove, T. Effect of elevated curing temperature on ceramsite concrete performance. *Constr. Build. Mater.* **2017**, *153*, 423–429. [CrossRef]
66. Chen, G.; Xu, P.; Mi, G.; Zhu, J. Compressive strength and cracking of composite concrete in hot-humid environments based on microscopic quantitative analysis. *Constr. Build. Mater.* **2019**, *225*, 441–451. [CrossRef]
67. Wei, J.-X.; Yu, Q.-J.; Zeng, X.-X.; Bai, R.-Y. Fractal dimension of pore structure of concrete. *J. South China Univ. Technol.* **2007**, *2*, 121–124.
68. Jianbo, Y.; Jihui, T.; Zhongmin, Z.; Xiaoming, Y.; Shenli, X. Application of shotcrete and its permeability, pore structure and mechanical properties. *Qinghai Jiaotong Keji* **2019**, 114–117.
69. İlhami, D.; Merve, G.H.; Süleyman, G. Gamma ray and neutron shielding characteristics of polypropylene fiber-reinforced heavyweight concrete exposed to high temperatures. *Constr. Build. Mater.* **2020**, *257*, 119596.
70. Yang, W. Microstructure of ettringite formed under different curing conditions. *J. Chin. Electron Microsc. Soc.* **2000**, 523–524.
71. Shen, J.; Xu, Q. Characteristics of pore structure change and compressive strength reduction of concrete under elevated Temperatures. *Mater. Rep.* **2020**, *34*, 2046–2051.
72. Schiller, K.K. *Mechanical Properties of Non-Metalic Brittle Materials*; Butter-Worth: London, UK, 1958.
73. Chen, L.; Zhou, Z.; Liu, G.; Cui, X.; Dong, Q.; Cao, H. Effects of substrate materials and liner thickness on the adhesive strength of the novel thin spray-on liner. *Adv. Mech. Eng.* **2020**, *12*, 1687814020904574. [CrossRef]
74. Liu, J.; Mei, Y.; Xia, R. A new wetting mechanism based upon triple contact line pinning. *Langmuir ACS J. Surf. Colloids* **2011**, *27*, 196–200. [CrossRef] [PubMed]
75. Ma, Q.; Duan, Y.; Su, H.; Tang, Y.; Wang, J.; Li, X.; Ma, C.H. Analysis on impacts of different rock wall temperatures on bonding strength between surrounding rock and shotcrete. *Water Resour. Hydropower Eng.* **2015**, *46*, 62–65.
76. Ou, Z. *Study on Performance and Technology of Shotcrete in Thermal Environment*; Southwest Jiaotong University: Chengdu, China, 2011.
77. Su, H.; Li, X.; Wang, J. Bonding strength of Shotcrete for tunnel under high ground temperature and its finite element analysis. *Water Resour. Hydropower Eng.* **2016**, *47*, 54–57.
78. Ma, F.; Su, H.; Ma, C.; Huang, S.; Li, Q. Experimental Simulation Study on the High Geothermal Temperature Surrounding Rock and Tunnel Lining Model Based on ANSYS. *Yellow River* **2018**, *40*, 113–116.
79. Su, H.; Huang, S.; Qu, C.-L. Analysis the distribution characteristics of pore structure in shotcrete affected by high temperature. *Sci. Technol. Eng.* **2016**, *16*, 225–229.
80. Fan, L.; Li, P.; Yu, Y.; Zhang, Z. Experimental study on the effect of curing temperature on shotcrete properties. *Bull. Chin. Ceram. Soc.* **2017**, *36*, 3487–3492.
81. Duan, L.; Zhang, Y.; Lai, J. Influence of ground temperature on shotcrete-to-rock adhesion in tunnels. *Adv. Mater. Sci. Eng.* **2019**, *2019*, 8709087. [CrossRef]
82. Tang, Y.; Xu, G.; Lian, J.; Su, H.; Qu, C. Effect of temperature and humidity on the adhesion strength and damage mechanism of shotcrete-surrounded rock. *Constr. Build. Mater.* **2016**, *124*, 1109–1119. [CrossRef]
83. Hassanein, A.Y.A.; Mohamed, N.; Farghaly, A.; Benmokrane, B. Effect of boundary element confinement configuration on the performance of GFRP-Reinforced concrete shear walls. *Eng. Struct.* **2020**, *225*, 111262. [CrossRef]
84. Torelli, G.; Gillie, M.; Mandal, P.; Draup, J.; Tran, V.X. A moisture-dependent thermomechanical constitutive model for concrete subjected to transient high temperatures. *Eng. Struct.* **2020**, *210*, 110170. [CrossRef]
85. Tang, X.; Wang, J.; Dong, C.; Wang, Y. Shear Strength of cementation plane between shotcrete and granite under high and variable temperature. *J. China Railw. Soc.* **2017**, *39*, 131–136.
86. Jianjun, T.; Murat, K.; Mingnian, W.; Congyu, D.; Xinghua, T. Shear strength characteristics of shotcrete–rock interface for a tunnel driven in high rock temperature environment. *Geomech. Geophys. Geo-Energy Geo-Resour.* **2016**, *2*, 331–341.
87. Saidi, M.; Vu, X.H.; Ferrier, E. Experimental and analytical analysis of the effect of water content on the thermomechanical behaviour of glass textile reinforced concrete at elevated temperatures. *Cem. Concr. Compos.* **2020**, *112*, 103690. [CrossRef]
88. Pan, P.; Tao, Z.; Lu, Y.; Chen, L.; Zeng, Y.; Li, D.; Pan, Z. Experimental study on high temperature mechanical properties of machine-made sand concrete based on residual strength. *Constr. Technol.* **2020**, *49*, 67–69.
89. He, M.; Guo, P.; Chen, X.; Meng, L.; Zhu, Y. Research on characteristics of high-temperature and control of heat-harm of sanhejian coal mine. *Chin. J. Rock Mech. Eng.* **2010**, *29*, 2593–2597.
90. Yu, C. *Research on Local Environment Cooling Technology of High Temperature Mine*; Shandong University of Science and Technology: Qingdao, China, 2017.

91. Twort, C.T.; Lowndes, I.S.; Pickering, S.J. An application of thermal energy analysis to the development of mine cooling systems. *J. Mech. Eng. Sci.* **2002**, *216*, 845–857. [CrossRef]
92. Jin, Q.; Li, H.; Chen, W. Experimental research on spray cooling on the permeability of concrete at different temperatures. *China South. Agric. Mach.* **2019**, *50*, 226–227.
93. Luo, Y.; Wang, H.; Zhang, Y. Research and application of cooling technology in roadway driving in the mine at high temperature. *Nonferrous Met. Sci. Eng.* **2020**, *11*, 85–91.
94. Xu, L. *Experimental Study on Cooling of Liquid CO$_2$ in High Temperature Tunneling Face*; Liaoning University of Engineering and Technology: Shenyang, China, 2017.
95. Liu, L.X.; Lv, L.; Liu, Z.; Wang, H.Y. Investigation on the Mechanical Behavior of Concrete at and after Elevated Temperature. *Build. Sci.* **2005**, *3*, 16–20.
96. He, Z. *Research on Test and Application of Thermal Insulation Shotcrete in High Temperature Roadway*; Anhui University of Science and Technology: Huainan, China, 2018.
97. Yunus, N.Z.M.; Ayub, A.; Wahid, M.A.; Satar, M.K.I.M.; Abudllah, R.A.; Yaacob, H.; Hassan, S.A.; Hezmi, M.A. Strength behaviour of kaolin treated by demolished concrete materials. In Proceedings of the IOP Conference Series: Earth and Environmental Science, Johor, Malaysia, 27–28 August 2019; Volume 220.
98. Zhou, J.; Li, H.; Zheng, W.; Ma, X.; Zhang, J.; Yang, Y. Effect of fly ash and metakaolin mixture on mechanical properties of high strength concrete at different temperature. *Bull. Chin. Ceram. Soc.* **2019**, *38*, 3952–3958.
99. Yang, T.; Liu, Z.; Yang, Y.; Wu, C. Experimental investigation on behavior of ultra-high performance concrete after high temperature. *J. Civ. Environ. Eng.* **2020**, *42*, 115–126.
100. Liu, T.; Liu, L.; Ji, S. Performance analysis of thermal insulation expanded perlite shotcrete in deep mine. *Ind. Constr.* **2018**, 739–741.
101. Pang, J.-Y.; Yao, W.-J.; Yao, W.-J. Experimental research on performance of new type of thermal insulation concrete material in high-temperature tunnel. *China Concr. Cem. Prod.* **2016**, *1*, 5–9.
102. Wang, R. Crack control of super-long concrete structure under temperature stress. *Henan Build. Mater.* **2020**, 111–112.
103. Lei, W. Research on the influence of cement temperature on concrete performance. *J. Commer. Concr.* **2019**, 40–42, 56.
104. Jiang, P.; Fang, J.; Pang, J.; Huang, J. Analysis of mechanical properties and microscopic properties of plant fiber sprayed concrete [J/OL]. *J. Yangtze River Sci. Res. Inst.* **2020**, 137–141, 149.
105. Lianjun, C.; Xixin, Z.; Guoming, L. Analysis of dynamic mechanical properties of sprayed fiber-reinforced concrete based on the energy conversion principle. *Constr. Build. Mater.* **2020**, *254*, 119167.
106. Yao, W.-J. *Research and Application of Insulated Concrete Shotcrete Support Technology for Deep High Ground Temperature Rock Roadway*; Anhui University of Science and Technology: Huainan, China, 2019.
107. Cui, S.; Liu, P.; Cui, E.; Su, J.; Huang, B. Experimental study on mechanical property and pore structure of concrete for shotcrete use in a hot-dry environment of high geothermal tunnels. *Constr. Build. Mater.* **2018**, *173*, 124–135. [CrossRef]
108. Cui, S.; Xu, D.; Liu, P.; Ye, Y. Exploratory study on improving bond strength of shotcrete in hot and dry environments of high geothermal tunnels. *KSCE J. Civ. Eng.* **2016**, *21*, 2245. [CrossRef]
109. Chu, Z.; Zhou, G.; Bi, S. Meso-characterization of the effective thermal conductivity of selected typical geomaterials in an underground coal mine. *Energy Explor. Exploit.* **2018**, *36*, 488–508. [CrossRef]
110. Bamonte, P.; Gambarova, P.G.; Nafarieh, A. High-temperature behavior of structural and non-structural shotcretes. *Cem. Concr. Compos.* **2016**, *73*, 42–53. [CrossRef]
111. Mu, W. *Quantitative Characterization of Interfacial Microstructure and Its Relationship with Mechanical Properties of Concrete*; Changjiang River Scientific Research Institute: Wuhan, China, 2017.
112. Gorlenko, N.P.; Sarkisov, Y.S.; Demyanenko, O.V.; Kopanitsa, N.O.; Sorokina, E.A.; Gorynin, G.L.; Nichinskiy, A.N. Fine-grained concrete fibre-reinforced by secondary mineral wool raw material. *J. Phys.* **2018**, *1118*, 012059. [CrossRef]
113. Benarchid, Y.; Taha, Y.; Argane, R.; Tagnit-Hamou, A.; Benzaazoua, M. Concrete containing low-sulphide waste rocks as fine and coarse aggregates: Preliminary assessment of materials. *J. Clean. Prod.* **2019**, *221*, 419–429. [CrossRef]
114. Monte, F.L.; Felicetti, R.; Miah, M.J. The influence of pore pressure on fracture behaviour of Normal-Strength and High-Performance Concretes at high temperature. *Cem. Concr. Compos.* **2019**, *104*, 103388. [CrossRef]

applied sciences

MDPI

Article

The Performance Modeling of Modified Asbuton and Polyethylene Terephthalate (PET) Mixture Using Response Surface Methodology (RSM)

Franky E. P. Lapian, M. Isran Ramli *, Mubassirang Pasra and Ardy Arsyad

Department of Civil Engineering, Faculty of Engineering, Hasanuddin University, Makassar 90245, Indonesia; lapianedwin@gmail.com (F.E.P.L.); mubapasra@gmail.com (M.P.); ardyarsyad@gmail.com (A.A.)
* Correspondence: isranramli@unhas.ac.id

Abstract: We often use the plastics daily, containing of polyethylene plastic polymers which recently can be utilized as additional material for road pavements. Several studies have attempted to find the optimum proportion of an asphalt mixture using modified Asbuton which is local bitumen abundantly deposited in Buton Island Indonesia, added with plastic waste. The optimum proportion of the asphalt mixture is influenced by many factors, such as the interactions of the material component in the asphalt mixture. To obtain the optimum proportion based a single factor, many studies employ statistical methods. This study aims to determine the optimum proportion for the asphalt mixture of the modified Asbuton with PET plastic waste by using a Response Surface Methodology (RSM). The employed RSM is the Expert Version 12 design (Stat-Ease, Inc., Minneapolis, MN, USA, 2020), in which the statistical modeling based on Box Behnken Design (BBD) and three factorial levels. The results obtained in this study show that the RSM optimization could achieve the asphalt mixtures characteristics including the stability, Marshall Quotient (MQ), Void in MIX (VIM), Void Mineral Aggregate (VMA) and density, in the level of satisfying the specification requirements of Ministry of Public Works of Indonesia. The optimum stability is at 2002.72 kg, fulfilled the minimum density of 800 kg. For the MQ, the optimal point of MQ is 500.68 kg/mm, satisfied the minimum the MQ standard minimum of 250 kg/mm. In addition, the optimal VIM is at 3.40%, satisfying the VIM specifications in the range of 3–5%. The optimal VMA response is at 21.65%, which is also satisfied the VMA specification, 15%.

Keywords: response surface methodology (RSM); PET plastic waste; modified asbuton

Citation: Lapian, F.E.P.; Ramli, M.I.; Pasra, M.; Arsyad, A. The Performance Modeling of Modified Asbuton and Polyethylene Terephthalate (PET) Mixture Using Response Surface Methodology (RSM). *Appl. Sci.* **2021**, *11*, 6144. https://doi.org/10.3390/app11136144

Academic Editor: Guoming Liu

Received: 21 March 2021
Accepted: 28 June 2021
Published: 1 July 2021

Publisher's Note: MDPI stays neutral with regard to jurisdictional claims in published maps and institutional affiliations.

1. Introduction

Pavement surface layer must have the ability to be a wearing layer and have good performance during its service life. The increase of traffic congestion has caused damage in pavement surface layer so that it cannot reach the expected service life. The repetition of traffic loads resulting from traffic density causes the accumulation of permanent deformation in the asphalt concrete mixture and decreases its service life [1,2]. One way to solve the problem is by using additives in the asphalt mixture. One of the additives is plastic waste, which contains polymer and found as plastomeric in the nature [3,4]. Several studies have suggested that PET plastic waste as an added material can improve the asphalt mixture performances [5–9].

In Indonesia, particularly in the island of Buton, the province of Southeast Sulawesi, natural deposit of asphalt or rock asphalt, namely Asbuton (Natural Asphalt Buton) can be found with abundant quantity. Asbuton is a naturally categorized as hydrocarbon material [10–13]. Asbuton bitumen content varies from 10% to 40%, and the rest is a mineral. The Asbuton deposit is quite large, around 600 million tons [14]. Moreover, the Asbuton deposit is estimated to be equivalent to 24 million petroleum asphalt [15–17]. It has been established in some previous studies that the combination of PET plastic waste and Asbuton can increase the asphalt mixture's stiffness, particularly the Marshall characteristics. They can improve several essential aspects of the asphalt mixture [18–20].

However, the optimum proportion of Asbuton and PET plastic waste for the asphalt mixture remains unclear. In general, experimental method was undertaken to evaluate a factor's effect in one experiment, associated with several variations and several experiments. In research terms, this is called a single factor experiment. The experimental method's weakness is that the conclusions obtained are only related to the experimental factors and are limited to 1 to 2 variables. Meanwhile, in reality, the quality of a product under study is influenced not only by one factor but also by several factors such as the level of modified Asbuton and the level of plastic waste. The proportions of these ingredients have interactions with one another, which significantly affect the quality of the Asphalt concrete-wearing course (AC-WC) mixture produced [21,22].

The method of quantifying the optimum proportion of PET plastic waste and the modified Asbuton in the asphalt mixture of AC-WC is still insufficient. Therefore, this study aims to investigate that optimum proportion of the modified Asbuton and PET plastic waste by using statistical techniques of Response Surface Methodology (RSM). This statistical method can take the contribution of two or more factors in an experiment into account and estimate the interactions and relationships between the experimental factors [23–26].

2. Materials and Methods

2.1. Physical Properties of Aggregate

Tables 1–3 show the result of laboratory tests, i.e., the characteristics of fine aggregate (stone ash), coarse aggregate characteristics and characteristics of the filler. The coarse aggregate, stone ash and filler are required to fulfil the road material's specification according to the 2018 Bina Marga (Indonesian Ministry of Public Works) General Specifications requirement.

Table 1. Physical properties of stone ash.

No.	Properties	Results	Specification		Unit
			Min	Max	
1	Water Absorption	2.79		3.0	%
2	Bulk Specific Gravity	2.45	2.5		
	SSD Specific Gravity	2.52	2.5		
	Apparent Specific Gravity	2.63	2.5		
3	Sand Equivalent	89.66	50		%

Table 2. Physical properties of coarse aggregate.

No.	Properties	Results.	Specifications		Unit
			Min	Max	
1	Water absorption				
	Coarse aggregate 5–10 mm	2.07		3.0	%
	Coarse aggregate 1–2 cm	2.08		3.0	%
2	Density				
	Coarse aggregate 0.5–1 cm				
	Bulk Specific Gravity	2.62	2.5		
	SSD Specific Gravity	2.67	2.5		
	Apparent Specific Gravity	2.77	2.5		
	Coarse aggregate 1–2 cm				
	Bulk Specific Gravity	2.62	2.5		
	SSD Specific Gravity	2.68	2.5		
	Specific Gravity	2.77	2.5		

Table 2. *Cont.*

No.	Properties	Results.	Specifications		Unit
			Min	Max	
3	Artificial Flake Index				
	Coarse aggregate 0.5–1 cm	20.10		25	%
	Coarse aggregate 1–2 cm	9.38		25	%
4	Abrasion				
	Coarse aggregate 0.5–1 cm	25.72		40	%
	Coarse aggregate 1–2 cm	24.36		40	%

Table 3. Physical properties of filler.

No.	Properties	Results	Specification		Unit
			Min	Max	
1	Water Absorption	2.28		3.0	%
2	Bulk Specific Gravity	2.60	2.5		
	SSD Specific Gravity	2.65	2.5		
	Apparent Specific Gravity	2.76	2.5		
3	Sand Equivalent	69.57	50		%

2.2. Characteristics of Asbuton Modification

Table 4 shows the testing results of the modified Asbuton, which is the asphalt extracted from Buton's bitumen asphalt granular and added with petroleum bitumen. The results describe the modified Asbuton's characteristics. It can be seen that the modified Asbuton used in this study qualified the specifications required by the 2018 General Specifications of Bina Marga.

Table 4. Physical Properties of Asbuton Modification.

No.	Test	Results	Specification	
			Min	Max
1	Penetration before weight loss (mm)	78.6	60	79
2	Flabby point (°C)	52	48	58
3	Ductility at 25 °C, 5 cm/min (cm)	114	100	
4	Flash point (°C)	280	200	
5	Specific gravity	1.12	1	
6	Weight loss (%)	0.3		0.8
7	Penetration after weight loss (mm)	86	54	

2.3. Characteristics of PET Plastic Waste

The plastic bottle used is a type of PET (Polyethylene Terephthalate), one of the polyethylene types, namely polymer consisting of long chains of monomers ethylene (IUPAC: ethene). The structure of molecularethene C_2H_4 is–CH_2–CH_2–n. Two CH_2 united by double bonds, Polyethylene is formed through a process polymerization of ethene. Figure 1 shows a thin surface polyethylene.

Figure 1. Thin surface Polyethylene.

PET type plastic is a brown type plastic made from petroleum. Its mechanical properties are strong, slightly translucent, high flexibility and the surface is somewhat greasy. At a temperature of 60 °C, PET is very resistant to chemical compounds, with a specific gravity of 0.91–0.94 gr/cm^3. PET is also a type of low-density polyethylene produced by free radical polymerization at high temperature (200 °C) and high pressure, it can be melted at temperature of 260 °C.

2.4. Marshall Stability

The testing method for the asphalt mixture is Marshall equipment test refers to SNI 06-2489-1991. The quotient of stability and flow magnitude is an indicator of potential flexibility of the asphalt mixture to cracking, and the quotient is called as Marshall Quotient.

2.5. Response Surface Methodology (RSM)

Experimental asphalt mixture design utilized in this study is Box-Behnken Design (BBD) in which RSM is used to optimize the mixture design. The BBD is designed to form a combination of two techniques with incomplete block design by adding the center points or center runs to the plan. The center run (NC) is an experiment with the center point at (0, 0, ... , 0), and there are at least three center runs for various sums of the factor k. If there are three factors, then the BBD design amounts to 12 plus a center run as in the equation matrix two and can be described in Figure 2.

$$
D = \begin{bmatrix}
-1 & -1 & 0 \\
-1 & 1 & 0 \\
1 & -1 & 0 \\
1 & 1 & 0 \\
-1 & 0 & -1 \\
-1 & 0 & 1 \\
1 & 0 & -1 \\
1 & 0 & 1 \\
0 & -1 & -1 \\
0 & -1 & 1 \\
0 & 1 & -1 \\
0 & 1 & 1 \\
0 & 0 & 0
\end{bmatrix}
\tag{1}
$$

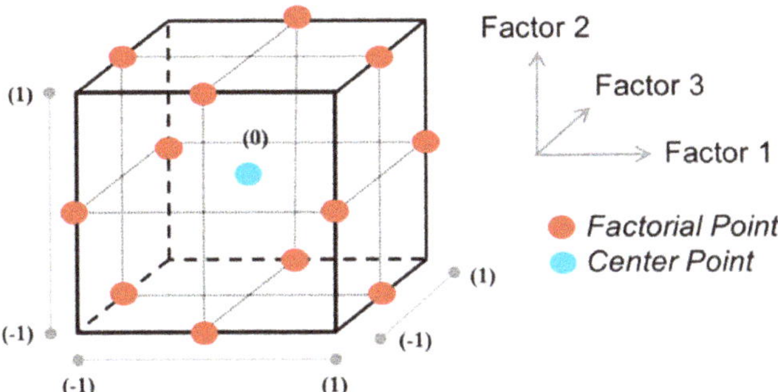

Figure 2. Box-Behnken Design.

The experimental mixture design was carried out using Response Surface Methodology (RSM) based on the Box-Behnken Design (BBD). The quadratic model and each variable vary on three levels. Design Expert Version 12 (Stat-Ease, Inc., Minneapolis, MN, USA) was used for regression analysis of experimental data and to plot response surfaces. The second stage uses the second-order quadratic polynomial equation to evaluate each independent variable's main effect and interaction on the response as given by Equation (2).

$$Y = \beta_0 + \sum_{i=1}^{n} \beta_i X_i + \sum_{i<j}^{n} \beta_{ij} X_i X_j + \sum_{j=1}^{n} \beta_{jj} X_j^2 \tag{2}$$

In Equation (2), Y represents the experimental response, i and j are linear and quadratic coefficients, respectively, β is the regression coefficient, n is the number of variables studied in the experiment, and X_i is a factor (independent variable). In this experiment, the independent variables (factor X) studied were X_1: the PET ratio to Asbuton, X_2: mixing temperature and X_3: mixing time, respectively. The response (Y) is characteristic of Marshall.

3. Results and Discussion

3.1. AC-WC Combined Aggregate Gradation

Figure 3 shows that the combined aggregate design or combined aggregate gradation made is within the standard specification according to the 2018 General Specifications of Bina Marga and has met the requirements for surface coating, so that the mixture design can be categorized as optimal mixture design.

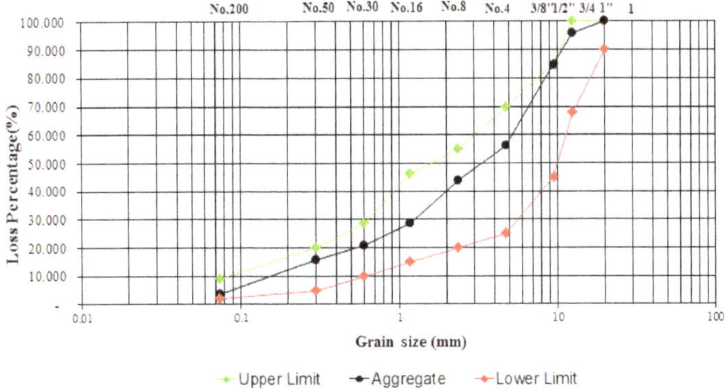

Figure 3. Combined aggregate gradation.

3.2. Mixture Design Results Based on RSM (Response Surface Methodology)

Table 5 shows the mixture design of asphalt mixture using Box Behnken Design (BBD) in the laboratory. In the fulfillment of the assumptions of Equation (2), we involved the five response variables in continuing at the modeling stage for optimization using the method Response Surface Methodology (RSM). This method is used to obtain the AC-WC asphalt mixture's optimization results using PET plastic waste and modified Asbuton based on Marshall characteristics (stability, MQ, VIM, VMA and density).

Table 5. Mixture design of asphalt mixture using BBD.

No	A: Asphalt Content (%)	B: PET Content (%)	C: Mixing Time (Minutes)	R1: Stability (kN)	R2: Flow (mm)	R3: MQ (kN/mm)	R4: VIM (%)	R5: VMA (%)	R6: VFB (%)	R7: Density
1	5.50	2.00	25.00	18.58	4.00	4.64	2.97	20.73	90.27	2296.00
2	5.50	4.00	25.00	15.35	4.00	3.84	1.77	20.68	90.70	2276.00
3	6.00	2.00	25.00	17.56	4.00	4.39	3.24	22.87	88.64	2278.00
4	6.00	4.00	25.00	17.83	4.00	4.46	5.00	20.37	90.93	2248.00
5	5.50	3.00	20.00	16.58	4.00	4.14	2.92	20.92	92.09	2203.00
6	5.50	3.00	30.00	16.97	4.00	4.24	2.78	20.86	90.72	2355.00
7	6.00	3.00	20.00	18.27	4.00	4.57	3.70	20.70	89.79	2196.00
8	6.00	3.00	30.00	18.82	4.00	4.70	5.25	20.48	84.27	2398.00
9	5.75	2.00	20.00	18.90	4.00	4.72	3.21	21.87	89.43	2234.00
10	5.75	2.00	30.00	19.76	4.00	4.94	2.98	21.78	89.23	2397.00
11	5.75	4.00	20.00	17.26	4.00	4.31	2.36	21.83	91.47	2264.00
12	5.75	4.00	30.00	17.53	4.00	4.38	2.57	20.98	87.49	2378.00
13	5.75	3.00	25.00	18.45	4.00	4.61	3.37	21.93	87.98	2263.00
14	5.75	3.00	25.00	19.75	4.00	4.94	3.95	22.35	85.78	2231.00
15	5.75	3.0	25.00	19.25	4.00	4.81	3.87	21.15	90.85	2316.00

The step of determining the optimum points simultaneously with RSM is through in 2 ways: manual experimentation on 15 combinations of the three parameters and by calculation using the RSM program. The first step is to determine which parameters are representing as X1, X2 and X3. Usually, in RSM, time and temperature are chosen as X1 and X2. Simultaneously, other parameters are expressed as X3 because, in this system, X1 also described as is the ratio of PET to Asbuton, the mixing temperature is X2 and X3 is the mixing time. Based on the RSM results, Figures 4–8 present the equations for predicting Marshall characteristics.

a.   Surface Stability Response Plots

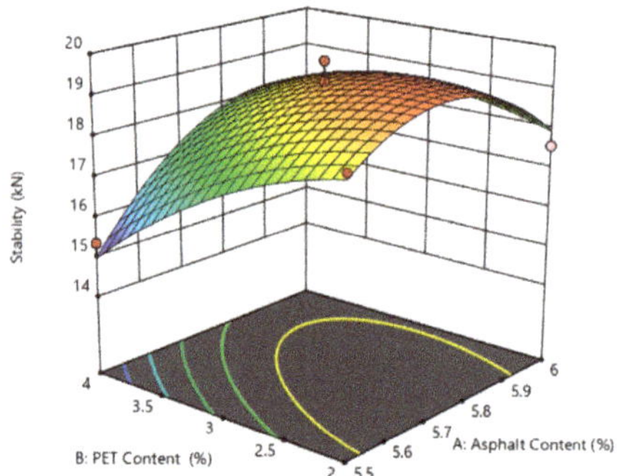

(**a**) Effect of asphalt content (A) and PET content (B).

Figure 4. *Cont.*

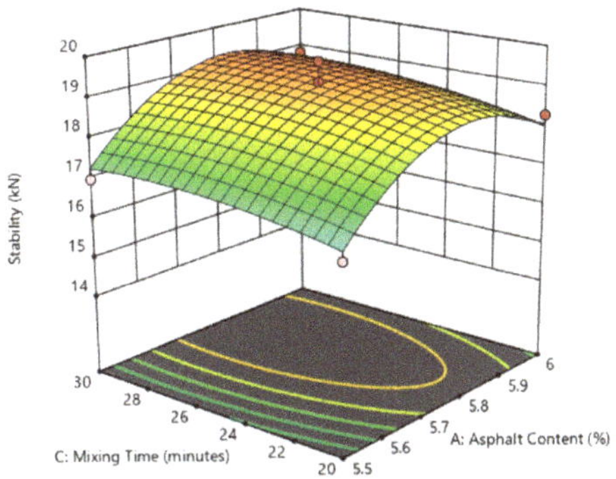

(**b**) Effect of asphalt content (A) and mixing time (C).

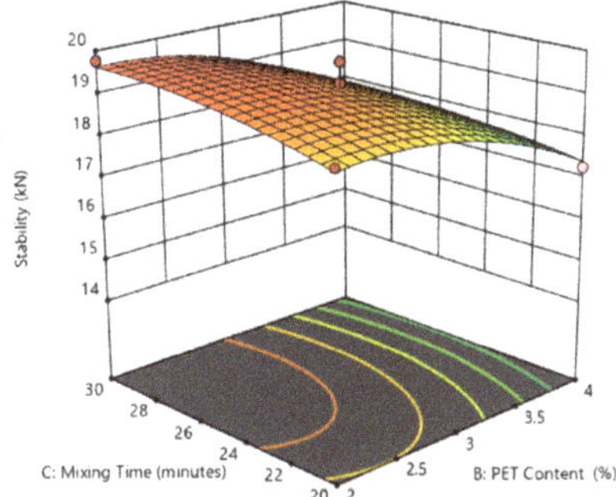

(**c**) Effect of PET content (B) and mixing time (C).

Figure 4. Contour and 3D surface response plots for stability.

b.　　Marshall Quetiont (MQ) Surface Response Plot

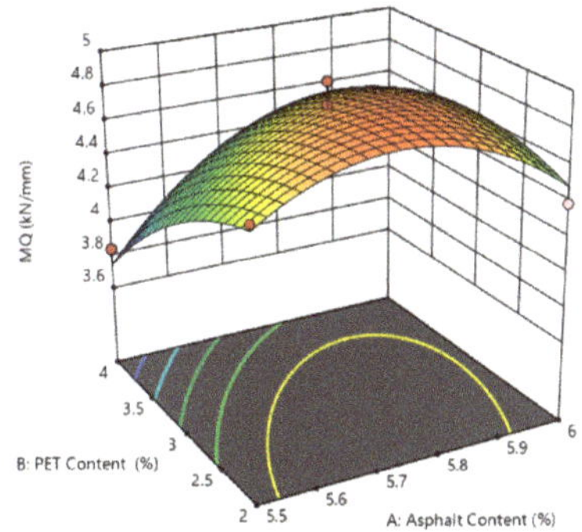

(**a**) Effect of asphalt content (A) and PET content (B).

Figure 5. *Cont.*

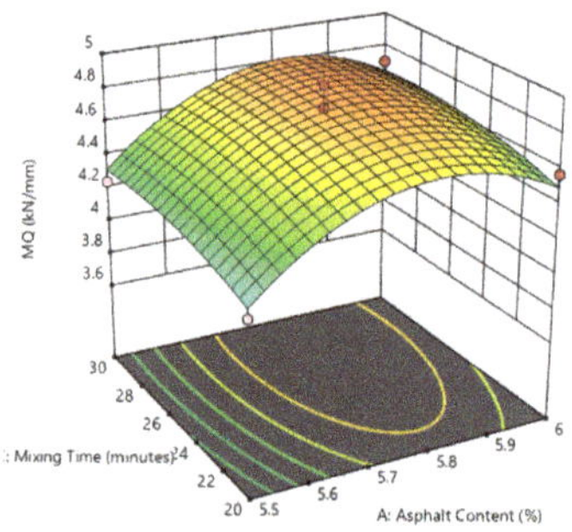

(**b**) Effect of asphalt content (A) and mixing time (C).

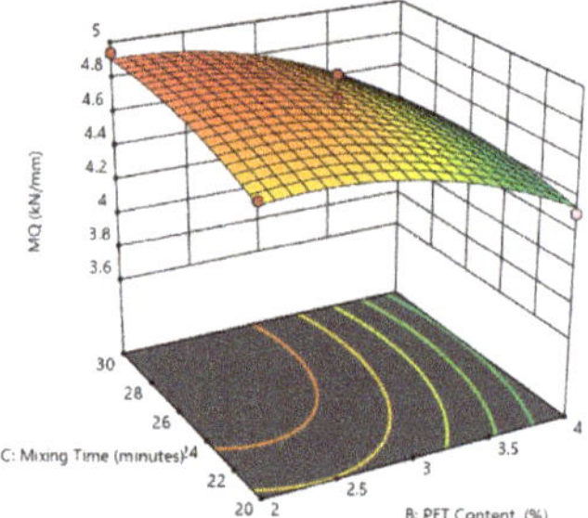

(**c**) Effect of PET content (B) and mixing time (C).

Figure 5. Contour and 3D surface response plots for MQ.

c.   VIM Surface Response Plots

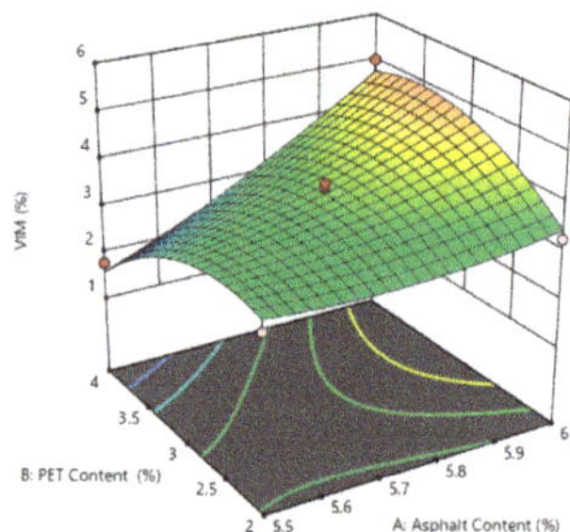

(**a**) Effect of asphalt content (A) and PET content (B).

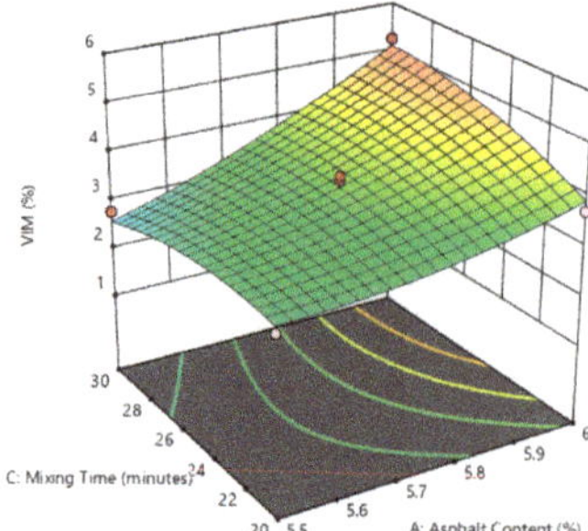

(**b**) Effect of asphalt content (A) and mixing time (C).

Figure 6. *Cont.*

(**c**) Effect of PET content (B) and mixing time (C).

Figure 6. Contour and 3D surface response plots for VIM.

d.   VMA Surface Response Plots

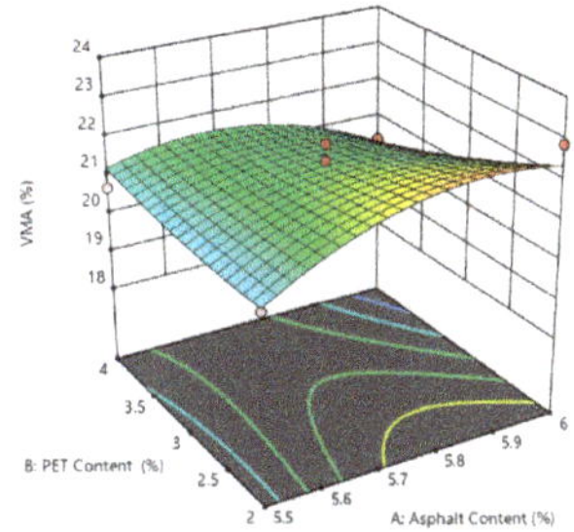

(**a**) Effect of asphalt content (A) and PET content (B).

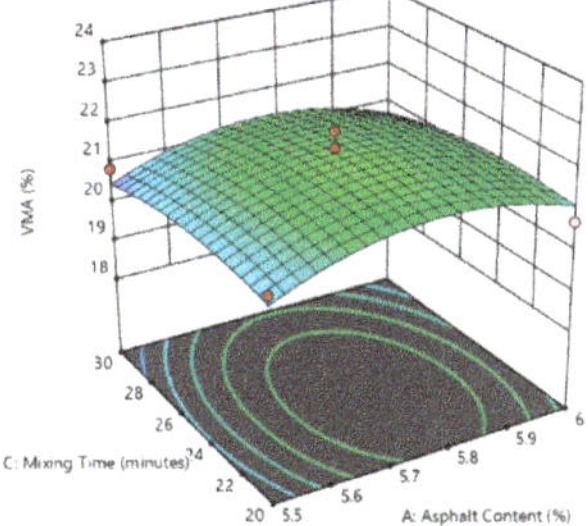

(**b**) Effect of asphalt content (A) and mixing time (C).

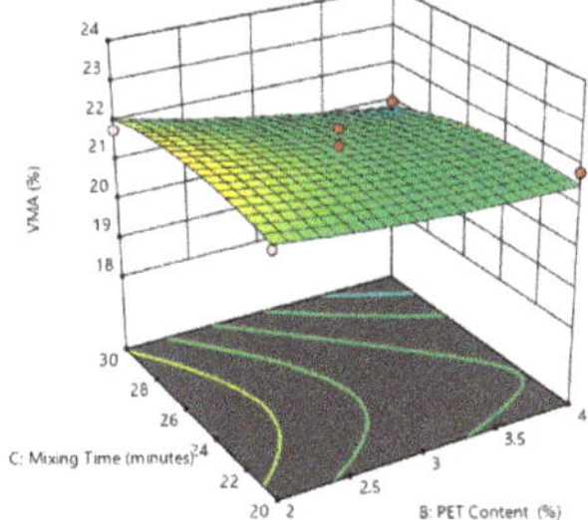

(**c**) Effect of PET content (B) and mixing time (C).

Figure 7. Contour and 3D surface response plots for VMA.

e.   Density Surface Response Plots

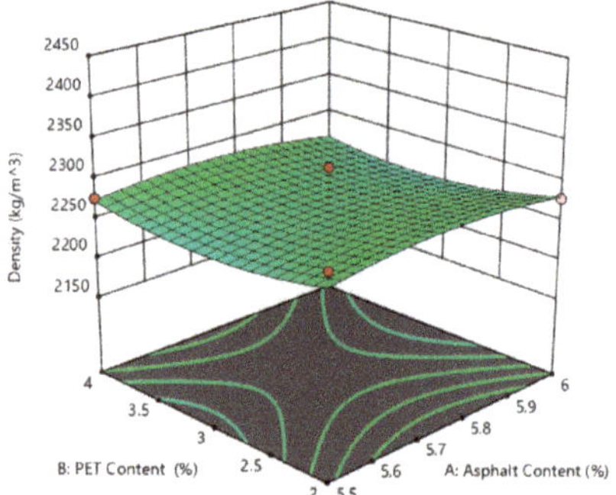

(**a**) Effect of asphalt content (A) and PET content (B).

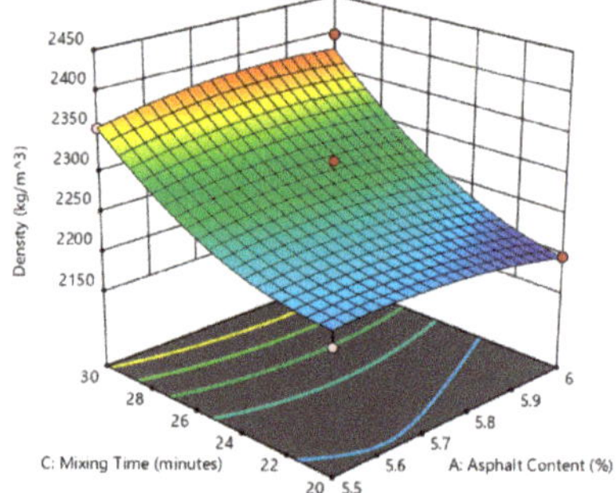

(**b**) Effect of asphalt content (A) and mixing time (C).

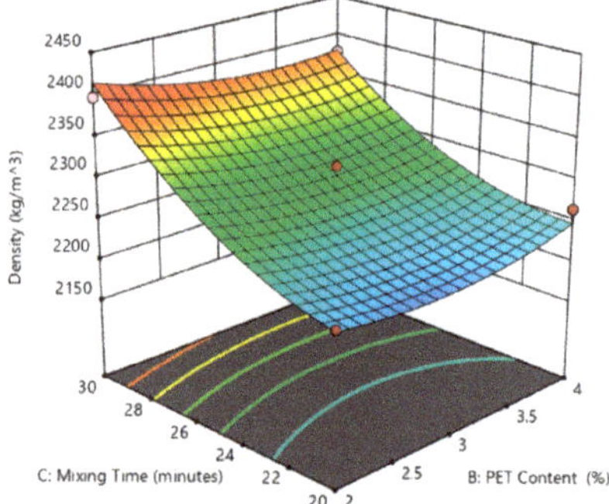

(**c**) Effect of PET content (B) and mixing time (C).

Figure 8. Contour and 3D surface response plot for density.

The ANOVA is shown in Table 6, and it can be seen that the order 2 which shows the order 2 model is suitable for this equation as evidenced by the f-value < f-table (19.16) for each variable. The f-table value with df lack of fit as df1 and df pure error as df 2 at alpha 0.05, the f-table is 19.16. If f-value > f-table then H_0 is stuck, where the assumption for lack of fit is that H_0 does not have a lack of fit and vice versa for H_1.

Table 6. ANOVA for predicting Marshall Stability based on RSM.

Variabel	Source	Sum of Squares	df	Mean Square	F-Value
Stability	Residual	1.51	5	0.3018	
	Lack of Fit	0.65	3	0.2163	0.50
	Pure Error	0.86	2	0.4300	
MQ	Residual	0.10	5	0.0193	
	Lack of Fit	0.04	3	0.0138	0.50
	Pure Error	0.05	2	0.0276	
VIM	Residual	0.88	5	0.1750	
	Lack of Fit	0.68	3	0.2258	2.29
	Pure Error	0.20	2	0.0988	
VMA	Residual	1.90	5	0.3799	
	Lack of Fit	1.16	3	0.3860	1.04
	Pure Error	0.74	2	0.3708	
Densitas	Residual	5732.75	5	1146.55	
	Lack of Fit	2046.75	3	682.25	0.37
	Pure Error	3686.00	2	1843.00	

The step of determining the optimum points simultaneously with RSM is carried out in two ways, namely by manual experimentation on 15 combinations of the three parameters and by calculation with the use of the RSM program. The first step to take is to determine which parameters are represented as X1, X2 and X3. In RSM, time and temperature are chosen as X1 and X2, while other parameters are expressed as X3 because in this system it is also expressed as X1 is the ratio of PET to Asbuton, the mixing temperature is X2 and X3 is the mixing time. Based on the RSM results, the equations for predicting Marshall characteristics are shown in Table 7.

Table 7. Equations for predicting Marshall Stability based on RSM.

No	Marshall Characteristics	The Equation for the Results of RSM	Adj. R^2
1	Stability	$19.15 + 0.63A - 0.85B + 0.26C + 0.88AB + 0.04AC - 0.15BC - 1.26A^2 - 0.56B^2 - 0.23C^2$	0.7975
2	MQ	$4.79 + 0.16A - 0.21B + 0.06C + 0.22AB + 0.0075AC - 0.04BC - 0.32A^2 - 0.14B^2 - 0.06C^2$	0.7921
3	VIM	$3.73 + 0.84A - 0.09B + 0.17C + 0.74AB + 0.42AC - 0.04BC - 0.32A^2 - 0.14B^2 - 0.06C^2$	0.7956
4	VMA	$3.73 + 0.84A - 0.09B + 0.17C + 0.74AB + 0.42AC - 0.04BC - 0.32A^2 - 0.14B^2 - 0.06C^2$	0.3241
5	Density	$2270 - 1.25A - 4.88B + 78.88C - 2.50AB + 12.50AC - 12.25BC - 12.87A^2 + 17.38B^2 + 30.87C^2$	0.7424

Note: The coefficients A, B, C refer to the linear response. AB, AC and BC are interactions between independent variables. A^2, B^2 and C^2 is a quadratic response involved in the process.

Based on the experimental design, the obtained VFB was in the range of 84.27–92.83%. However, this data can only reach order 1, as evidenced by the test's lack of fit in model 1, the decision to fail to reject H_0, which means that the model is suitable, or there is no lack of a fit model. This result leaves no increase to the 2nd order in this model. While in RSM, optimization will occur in order 2.

3.3. Optimization of PET Levels in Marshall Characteristics

Regarding the purpose of the study, Table 8 shows the optimizing results for PET content as much as possible in Marshall characteristics. It is shown that the optimal PET content used was at 3.84%, with a stationary point of 0.844. It is noted that the mixing time could not affect the stability and density of the asphalt mixture since the PET is not melted but crystalized in temperature of AC-WC mixture.

Table 8. Minimum PET content.

Response Variable	Optimal Response	Optimal A: Asphalt Content (%)	Optimal B: PET Content (%)	Optimal C: Mixing Time (Minutes)
Stability	19.64 kN	5.73	2.07	29.29
MQ	4.91 kN/mm	5.76	3.46	22.85
VIM	3.40%	5.56	2.49	23.07
VMA	21.65%	5.69	3.84	22.54
Density	2223.06 kg/m^3	5.40	2.67	18.89

4. Conclusions

The present study has observed the seven components of Marshall characteristics. There are two components that RSM cannot optimize, i.e., Flow and Void Filled Bitumen (VFB). Statistical tests cannot carry out the response flow because it does not have data diversity and the VFB response matches the 1st order model. This result may be due to the data range that is too small. Asphalt levels, PET plastic waste levels and mixing time have different optimum points for each response of the AC-WC mixture using modified Asbuton as the binder.

The RSM analysis results showed that optimum proportion of asphalt and PET contents in the AC-WC mixture could achieve the values of stability and MQ, VIM, VMA and density, meet with the technical specifications required by Indonesia Ministry of Public Works. The results also show that the PET content could enhance the VIM in the AC-WC mixtures, indicating the durability of the AC-WC mixtures against the water infiltration. The findings suggested the PET and the modified Asbuton for AC-WC asphalt mixtures would be potential for future application as environmentally friendly materials for asphalt pavement technology.

Author Contributions: Conceptualization, F.E.P.L. and M.I.R.; Data curation, F.E.P.L.; Formal analysis, A.A.; Funding acquisition, M.P.; Methodology, M.I.R. and A.A.; Resources, M.P.; Software, F.E.P.L. and A.A.; Supervision, M.I.R., M.P. and A.A.; Validation, M.P.; Visualization, F.E.P.L.; Writing—original draft, F.E.P.L.; Writing—review & editing, M.I.R. All authors have read and agreed to the published version of the manuscript.

Funding: This research received no external funding.

Institutional Review Board Statement: Not applicable.

Informed Consent Statement: Not applicable.

Data Availability Statement: The data presented in this study are available on request from the corresponding author. The data are not publicly available due to restrictions eg privacy or ethical.

Conflicts of Interest: The authors declare no conflict of interest.

References

1. Tayfur, S.; Ozen, H.; Aksoy, A. Investigation of Rutting Performance of Asphalt Mixtures Containing Polymer Modifiers. *Constr. Build. Mater.* **2005**, *21*, 328–337. [CrossRef]
2. Birgisson, B.; Montepara, A.; Romeo, E.; Roncella, R.; Napier, J.A.L.; Tebaldi, G. Determination and Prediction of Crack Patterns in Hot Mix Asphalt (HMA) Mixtures. *Constr. Build. Mater.* **2008**, 664–673. [CrossRef]
3. Ahmadinia, E.; Zargar, M.; Karim, M.R.; Abdelaziz, M.; Shafigh, P. Using waste plastic bottles as an additive for stone mastic asphalt. *Mater. Des.* **2011**, *32*, 4844–4849. [CrossRef]
4. Ahmadinia, E.; Zargar, M.; Karim, M.R.; Abdelaziz, M.; Ahmadinia, E. Performance evaluation of utilization of waste Polyethylene Terephthalate (PET) in stone mastic asphalt. *Constr. Build. Mater.* **2012**, *36*, 984–989. [CrossRef]
5. Musa, E.I.A.; Haron, H.E.F. Effect of the Low Density Polyethylene Carry Bags Waste on the Asphalt Mixture. *Int. J. Eng. Res. Sci. Technol.* **2014**, *3*, 86–93.
6. Rajput, P.S.; Yadav, R.K. Use of Plastic Waste in Bituminous Road Construction. *Int. J. Sci. Technol. Eng. (Ijste)* **2016**, *2*, 509–513.
7. Sojobi, A.O.; Nwobodo, S.E.; Aladegboye, O.J.; Pratico, F.G. Recycling of Polypthylene Terephthalate (PET) Plastic Bottle Wastes in Bituminous Asphaltic Concrete. *Cogent Eng.* **2016**. [CrossRef]

8. Soltani, M.; Moghaddam, T.B.; Karim, M.R.; Baaj, H. Analysis of Fatigue Properties of Unmodified and Polyethylene Terephthalate Modified Asphalt Mixtures Using Response Surface Methodology. *Eng. Fail. Anal.* **2015**, *58*, 238–248. [CrossRef]
9. Gaus, A.; Tjaronge, M.W.; Ali, N.; Djamaluddin, R. Compressive Strength of Asphalt Concrete Binder Course (AC-BC) Mixture Using Buton Granular Asphalt (BGA). In Proceedings of the 5th International Conference of Euro Asia Civil Engineering Forum (EACEF-5), Surabaya, Indonesia, 15–18 September 2015; Volume 125, pp. 657–662.
10. Tjaronge, M.W.; Irmawaty, R. Influence of Water Immersion on Physical Properties of Porous Asphalt Containing Liquid Asbuton as Bituminous Asphalt Binder. In Proceedings of the 3rd International Conference and Sustainable Construction Material and Technologies-SCTM, Kyoto, Japan, 18–22 August 2003.
11. Tumpu, M.; Tjaronge, M.W.; Djamaluddin, A.R.; Amiruddin, A.A.; One, L. Effect of limestone and buton granular asphalt (BGA) on density of asphalt concrete wearing course (AC-WC) mixture. *IOP Conf. Ser. Earth Environ. Sci.* **2020**, *419*, 012029. [CrossRef]
12. Tjaronge, M.W.M.; Djamaluddin, A.R. Prediction of long-term volumetric parameters of asphalt concrete binder course mixture using artificial aging test. *IOP Conf. Ser. Earth Environ. Sci.* **2020**, *419*, 012058. [CrossRef]
13. Suryana, A. *Inventory on Solid Bitumen Sediment Using 'Outcrop Drilling' in Southern Buton Region, Buton Regency, Province Southeast Sulawesi, Colloquium on Result Activities of Mineral Resources Inventory*; Directorate Minerals: Bandung, Indonesian, 2003.
14. Mabui, D.S.; Tjaronge, M.W.; Adisasmita, S.A.; Pasra, M. Resistance to cohesion loss in cantabro test on specimens of porous asphalt containing modified asbuton. *IOP Conf. Ser. Earth Environ. Sci.* **2020**, *419*, 012100. [CrossRef]
15. Iroth, M.W.; Tjaronge, M.W.; Pasra, M. Influence of short term oven aging on volumetric properties of asphalt concrete mixture containing modified Buton asphalt and limestone powder filler. *IOP Conf. Ser. Earth Environ. Sci.* **2020**, *419*, 012072. [CrossRef]
16. Tjaronge, M.W.; Adisasmita, S.A.; Hustim, M. Analysis of stability of residue asphalt emulsion mixture containing Buton Granular Asphalt (BGA). *IOP Conf. Ser. Earth Environ. Sci.* **2020**, *419*, 012073. [CrossRef]
17. Mabui, D.S.; Tjaronge, M.W.; Adisasmita, S.A.; Pasra, M. Performance of porous asphalt containing modified Buton asphalt and plastic waste. *Int. J. Geomate* **2020**, *18*, 118–123. [CrossRef]
18. Ramli, M.I.; Pasra, M.; Amiruddin, A.A. The sustainable performance challenge of asphalt mixture using polypropylene due to environmental weather. *IOP Conf. Ser. Earth Environ. Sci.* **2020**, *419*, 012075. [CrossRef]
19. Meraudje, A.; Ramli, M.I.; Pasra, M.; Amiruddin, A.A. The potential utilization of Polyethylene Terephthalate (PET) waste as fine aggregate replacement in asphalt mixture. *IOP Conf. Ser. Earth Environ. Sci.* **2020**, *419*, 012036. [CrossRef]
20. Lapian, F.E.P.; Ramli, M.I.; Pasra, M.; Arsyad, A. Opportunity applying response surface methodology (RSM) for optimization of performing butonic asphalt mixture using plastic waste modifier: A preliminary study. *IOP Conf. Ser. Earth Environ. Sci.* **2020**, *419*, 012032. [CrossRef]
21. Chávez-Valencia, L.E.; Manzano-Ramírez, A.; Alonso-Guzmán, E.; Contreras-García, M.E. Modeling of the performance of asphalt pavement using response surface methodology—The kinetics of the aging. *Build. Environ.* **2007**, *42*, 933–939. [CrossRef]
22. Ghasemi, I.; Karrabi, M.; Mohammadi, M.; Azizi, H. Evaluating the effect of processing conditions and organoclay content on the properties of styrene-butadiene rubber/organoclay nanocomposites by response surface methodology. *Express Polym. Lett.* **2010**, *4*, 62–70. [CrossRef]
23. Nassar, A.I.; Thom, N.; Parry, T. Optimizing the mix design of cold bitumen emulsion mixtures using response surface methodology. *Constr. Build. Mater.* **2016**, *104*, 216–229. [CrossRef]
24. Mogahaddam, T.B.; Soltani, M.; Karim, M.R.; Baaj, H. Optimization of asphalt and modifier contents for polyethylene terephthalate modified asphalt mixtures using response surface methodology. *Measurement* **2015**, *74*, 159–169. [CrossRef]
25. *General Specifications of Indonesia 2018 Indonesia Requirement*; Director General of Bina Marga, Public Work Ministry of Indonesia: Jakarta, Indonesia, 2018. (In Indonesian)
26. Standard National of Indonesia. *Standard Test Method of Asphalt Mix with Marshall Test*; SNI 06-2489-1991; Centre of Transportation of Research and Development Board of Public Work Ministry of Indonesia: Bandung, Indonesia, 1991. (In Indonesian)

Article

Research on Compressive Strength of Manufactured Sand Concrete Based on Response Surface Methodology (RSM)

Hui Ma [1], Zhenjiao Sun [2] and Guanguo Ma [1],*

[1] College of Energy and Mining Engineering, Shandong University of Science and Technology, Qingdao 266590, China; hughie_ma@163.com
[2] College of Safety and Environmental Engineering, Shandong University of Science and Technology, Qingdao 266590, China; sunzhenjiao@foxmail.com
* Correspondence: ma156154682@126.com; Tel.: +86-1515-441-9905

Abstract: Traditional natural river sand is used as a fine aggregate for concrete, but due to the severe environmental situation in recent years, many places have asked for a ban or restriction on the extraction of river sand. This has resulted in an increasing demand for concrete using machine-made sand instead of natural sand. The estimation and prediction of the compressive strength of concrete is very important in civil engineering applications. In this investigation, a Box–Behnken test model was established to analyze the effect of stone powder (SP), pulverized fuel ash (PFA), and silica fume (SF) contents on the compressive strength of manufactured sand concrete using response surface methodology (RSM). A prediction model for the compressive strength of manufactured sand concrete was developed using multiple regression analysis with SP, PFA, and SF content as factors and compressive strength as the response value. In addition, the interaction of stone powder (SP), pulverized fuel ash (PFA), and silica fume (SF) content was analyzed according to the response surface and contour. The investigation showed that for single factors, SP had the greatest effect on the compressive strength of the manufactured sand concrete, with PFA having the second greatest effect, and SF the least; for the interactions, SP and PFA had the most significant effect, and the interaction between SP and SF and PFA and SF had the same effect on the compressive strength.

Keywords: manufactured sand concrete; RSM; SP; PFA; SF; compressive strength

Citation: Ma, H.; Sun, Z.; Ma, G. Research on Compressive Strength of Manufactured Sand Concrete Based on Response Surface Methodology (RSM). *Appl. Sci.* **2022**, *12*, 3506. https://doi.org/10.3390/app12073506

Academic Editor: Carlos Thomas

Received: 9 March 2022
Accepted: 29 March 2022
Published: 30 March 2022

1. Introduction

With the high speed of economic development and the increased investment in infrastructure in countries around the world, the amount of concrete used has increased dramatically [1,2]. Natural sand is a fundamental component of concrete, and with the increase in concrete consumption, more and more river sand is being mined. The widespread mining of river sand can have a negative impact on the surrounding environment; therefore, some rivers have now banned the mining of river sand. Manufactured sand is gradually replacing natural sand in building materials [3]. As shown in Figure 1, the consumption of aggregates and the proportion of manufactured sand used in China has increased from year to year over the last decade.

Manufactured sand is made from the crushing of sedimentary rocks and has some unique components not found in natural sand [4,5]. Manufactured sand is able to avoid the reactions between the active silica components of natural sand and the alkali metal hydroxides of cement. Numerous publications have shown that some of the properties of manufactured sand concrete are better than those of natural sand concrete [6–8]. Various mineral admixtures are an essential component to improve the performance of the manufactured sand concrete [9]. SP may induce hydride precipitation, which increases the strength of concrete by increasing the content of effective crystallization products [10]. PFA promotes deflocculation in the hydration of cement clinker to reduce water consumption and fills pores to prevent agglomeration between cement particles. The addition of

SF to concrete markedly improves the adhesion and cohesion of shotcrete and increases the sequential forming thickness. In addition, some publications have shown that the addition of SP, PFA, and SF to manufactured sand concrete can have a "superimposed effect" on each other, reducing the heat of hydration of the concrete while improving its mechanical properties [11–13]. Prakash [14], Beixing [15] et al. investigated the effect of SP content in manufactured sand on the mechanical properties of concrete. Skaropoulou [16], Schmidt [17] et al. showed that the SP content of manufactured sand has an important influence on the durability of concrete. Wentao et al. [18] investigated the effects of PFA alone, SP alone, and a combination of PFA and SP on the workability and strength of manufactured sand concrete. Heng et al. [19] modified the concrete by incorporating PFA into the manufactured sand. His design for shotcrete, with a PFA admixture of 40% and a water-cement ratio of 0.37, reduced the water consumption while reducing the rebound rate of the shotcrete. Jain [20] found that the addition of marble powder reduced the strength of ordinary Portland cement. After curing the concrete with 20% marble powder for 28 days, a maximum compressive strength of 54.5 MPa was achieved. Currently, most publications report the effect of a single admixture of SP, PFA, and SF or both on the strength of manufactured sand concrete. However, no investigation of the combined effect of the three on compressive strength has been reported.

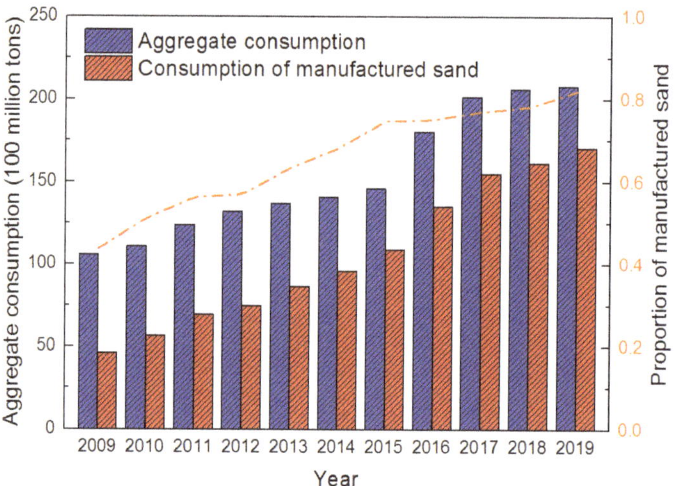

Figure 1. Aggregate consumption and proportion of manufactured sand in China.

Response surface methodology (RSM) is a statistical method for solving multivariate problems that enables experiments to be conducted using rational experimental design methods with multivariate quadratic equations to be fitted as a function of the relationship between factors and response values. The Box–Behnken experimental design in RSM has been widely used in engineering applications since it was proposed [21], and research on admixtures and concrete has recently become a hot topic. Zhang et al. [22] applied RSM to the design of a recycled aggregate permeable concrete mix and found a suitable combination of aggregate gradation and admixture mix. Natalia et al. [23] used the two-level central composite design in RSM to optimize the ratio of water-to-binder, PFA-to-binder and iron oxide nanoparticles-to-binder for Portland cement permeable concrete. Rajesh and Kumar [24] used Box–Behnken design optimization with RSM to obtain concrete with good hardening and functional properties. Khudhair et al. [25] used RSM to determine a model for predicting the compressive strength of high-performance concrete formulated by a high water reducing and setting accelerating superplasticizer as a function of the proportion of the constituents used.

A review of the literature reveals that the use of admixtures in manufactured sand concrete is common. The effect of various admixtures on the properties of manufactured sand concrete is an issue that needs to be addressed. Among these issues, the estimation and prediction of the compressive strength of manufactured sand is very important in civil engineering applications. At the same time, the RSM is able to fit the relationship between the factors and the response values obtained. Therefore, in this investigation, the Box–Behnken design based on RSM used SP, PFA, and SF admixtures as factors and compressive strength as response values to study the effect of the three admixtures on compressive strength. A multivariate predictive regression model for each factor was developed to analyze the magnitude of the effect of the factors. This investigation can provide an experimental basis and theoretical guidance for the design of manufactured sand concrete.

2. Experimental Materials and Methods

2.1. Experimental Materials

2.1.1. Cement and Water

In this work, ordinary silicate PO42.5 cement produced by Shandong Shanshui Cement Group Co., Ltd. (Rizhao, Shandong) was used for the experiments and the quality was in accordance with the GB175-2020 standard (General Purpose Silicate Cement in China) and ASTM C150 (Standard Specification for Portland Cement). The chemical composition of the cement is shown in Table 1. The specific surface area of the cement is 338 m^2/kg, the loss on burning is 4.54%, the initial setting time is greater than 45 min, and the final setting time is less than 600 min.

Table 1. Chemical composition of concrete.

Constituents	SiO$_2$	Al$_2$O$_3$	Fe$_2$O$_3$	CaO	MgO	SO$_3$
Content (wt%)	20.81	4.54	3.15	64.22	2	2.5

All water used for the experiments was from tap water, in accordance with the requirements of JGJ63-2006, Standard of Water for Concrete.

2.1.2. Manufactured Sand and Coarse Aggregates

In this work, all of the manufactured sand was obtained from the first phase of the Qingdao underground railway, Line 6, project in Shandong Province. The underground railway construction was carried out by blasting and the resulting stones were large and needed to be crushed and sieved before they could be used. In this work, a jaw crusher was used to crush the blocks and screen the manufactured sand according to the gradation as shown in Figure 2.

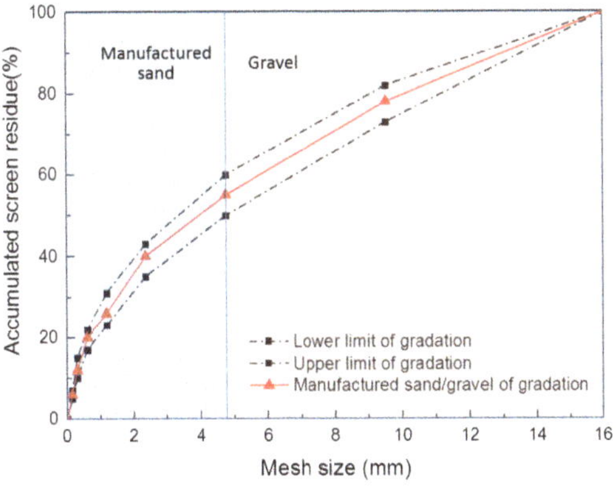

Figure 2. Manufactured sand and gravel gradation.

The coarse aggregate was made of durable gravel with a particle size of 5–10 mm, in accordance with GB50086-2015 Technical Specification for Anchor-Shotcrete Support. As shown in Figure 2, in this work both, the manufactured sand and gravel aggregate grade lines are located between the upper and lower lines of the shotcrete technical standard grade.

2.1.3. Stone Powder (SP)

The SP content within a reasonable range can be give concrete good workability; beyond this range, SP will have a negative impact. If the water-SP ratio is too large, it is easy to produce segregation and water secretion. For concrete with a large water to cement ratio, SP can be relied upon to reduce the water-SP ratio to improve cohesion and enhance water retention and reduce segregation and water secretion. In addition, if the SP content is too high, it will reduce the flowability of the concrete.

The SP in this work was obtained by sieving the crushed machine-made sand with a particle size of less than <0.075 mm. Three levels of SP content were set at 5%, 10%, and 15% of the cement mass, respectively.

2.1.4. Pulverized Fuel Ash (PFA)

PFA can act as an activator and filler, allowing the structural density of the concrete to increase. The morphological and micro-aggregate filling effects of PFA can improve the flowability of concrete mixtures in the early stages of concrete mix formation. The PFAs in the experiments in this work were all supplied by Class F, produced by Henan Hengyuan New Materials Co. (Zhengzhou, China) The chemical composition of PFA is shown in Table 2. The fineness, water requirement ratio, burn vector, and water content are 8.7%, 91%, 2.8%, and 0.2%, respectively, which meet the requirements of Grade I PFA for the relevant parameter index. The PFA content in this work was set at three levels of 10%, 15%, and 20% of the cement mass.

Table 2. Chemical composition of PFA.

Constituents	SiO_2	Al_2O_3	Fe_2O_3	CaO	TiO_2	K_2O	MgO
Content (wt%)	32.61	24.54	3.45	4.42	0.93	0.84	0.56

2.1.5. Silicon Fume (SF)

If the content of SF in the concrete is too small, the concrete performance is not much improved, but if the content is too much, the concrete is too sticky and hard to form, and the dry shrinkage deformation is large, showing poor frost resistance. The SF in this paper is produced by Henan Hengyuan New Materials Co. and its chemical composition is shown in Table 3. The density is 2.4 g/cm^3 and the specific surface area is 75,000 m^2/kg. Three levels of SF are set, 2.5%, 5%, and 7.5% of the cement mass, respectively.

Table 3. Chemical composition of SP.

Constituents	SiO_2	Al_2O_3	Fe_2O_3	CaO	SO_3	K_2O	MgO	Na_2O
Content (wt%)	92.5	0.2	0.6	0.12	0.44	1.52	0.15	0.37

2.2. Experimental Methods

2.2.1. Concrete Mixing Ratio

The water to cement ratio for the concrete in this work is 0.5, where the cement, gravel, manufactured sand, and water are configured in the ratio of 1:1.5:2.25:0.5. SP, PFA, and SF as a percentage of the mass of the cement. According to the Box–Behnken experiment design in the RSM, a total of 17 mix ratios are required at different contents, of which the SP, PFA, and SF contents are shown in Table 4 as a percentage of the cement mass. The test results are the average of the compressive strength of the three blocks.

Table 4. SP, PFA, and SF contents.

No.	SP	PFA	SF
1	5%	10%	5%
2	15%	10%	5%
3	5%	20%	5%
4	15%	20%	5%
5	5%	15%	2.5%
6	15%	15%	2.5%
7	5%	15%	7.5%
8	15%	15%	7.5%
9	10%	10%	2.5%
10	10%	20%	2.5%
11	10%	10%	7.5%
12	10%	20%	7.5%
13	10%	15%	5%
14	10%	15%	5%
15	10%	15%	5%
16	10%	15%	5%
17	10%	15%	5%

2.2.2. Concrete Block Making

Concrete blocks are made in accordance with the requirements of the Standard for Test Methods of Mechanical Properties on Ordinary Concrete GB/T50081-2016 standard for the production of specimen dimensions, and the mold size is $100 \times 100 \times 100$ mm. The concrete was prepared in accordance with the concrete ratios in Section 2.2.1 and the additive content in Table 4. The gravel, manufactured sand, and cement are first mixed in a concrete mixer for 1 min, after which water and other admixtures are added and mixed for an additional 3 min. The mixed concrete was poured into the molds and placed on a vibrating table for a period of 4 min. It should be noted that the vibration process resulted in a reduction of concrete in the molds; therefore, concrete had to be continuously added to the molds until it overflowed. Finally, the concrete was scraped off the molds, cured at room temperature for 24 h, and then demolded. After demolding, the concrete blocks were cured for 28 days according to the standards [26]. The concrete block-making process is shown in Figure 3.

Figure 3. Concrete block-making process.

2.2.3. Compressive Strength Test

The test of compressive strength is performed by a digital display type pressure tester (DYE-2000); the test content is evaluated for uniaxial compressive strength. The test procedure is shown in Figure 4. The cubic block is placed on the base of the pressure testing machine and the pressure is transmitted downwards to the concrete block through the upper plate. The loading speed of the force is hydraulically controlled at around 0.8 mm/min. The pressure tester records and outputs the maximum pressure value as the compressive strength [27].

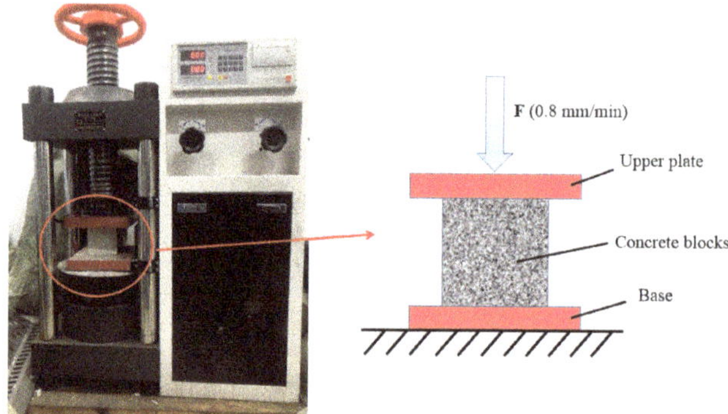

Figure 4. The test of compressive strength.

3. Results and Discussion

3.1. Box–Behnken Experiment Design and Significance Test

3.1.1. Box–Behnken Experiment Design

The Box–Behnken test factors and levels are shown in Table 5, using the uniaxial compressive strength values as the response values and the SP content (X_1), PFA content (X_2), and SF content (X_3) as the investigating factors. The test results and analysis are shown in Table 6.

Table 5. Factors and levels of Box–Behnken experiments.

Factors	−1	0	1
X_1 (SP content)	5%	10%	15%
X_2 (PFA content)	10%	15%	20%
X_3 (SF content)	2.5%	5%	7.5%

Table 6. Results and analysis of Box–Behnken experiments.

No.	X_1	X_2	X_3	Y/Compressive Strength (MPa)
1	5%	10%	5%	43
2	15%	10%	5%	45
3	5%	20%	5%	46
4	15%	20%	5%	44
5	5%	15%	2.5%	46
6	15%	15%	2.5%	41
7	5%	15%	7.5%	46
8	15%	15%	7.5%	44
9	10%	10%	2.5%	43
10	10%	20%	2.5%	38
11	10%	10%	7.5%	41
12	10%	20%	7.5%	39
13	10%	15%	5%	46
14	10%	15%	5%	48
15	10%	15%	5%	49
16	10%	15%	5%	47
17	10%	15%	5%	48

The least squares method was used to fit the experimental data, and a regression model was developed as follows:

$$Y = 47.6 - 0.87X_1 - 0.62X_2 + 0.25X_3 - X_1X_2 + 0.75X_1X_3 + 0.75X_2X_3 + 0.45X_1X_1 - 3.55X_2X_2 - 3.8X_3X_3 \tag{1}$$

$$R^2 = 0.85 \tag{2}$$

3.1.2. Significance Test

The standard quadratic regression equation (Equation (1)) was analyzed for variance, and the results are shown in Table 7. The model was tested for significance using ANOVA. The significance level was set at 0.05, i.e., when the p-value was less than 0.05, the indicator was considered significant; when the p-value was greater than 0.05, the indicator was considered insignificant. Table 7 shows that the p-value of the quadratic regression model for compressive strength is less than 0.05, and the multivariate correlation coefficient R^2 is 0.85. This indicates that the regression equation approximates the true surface well, and the model can accurately predict the compressive strength of the manufactured sand concrete.

Table 7. Variance analysis of response surface experiments results.

Source	Squares	df	Square	Value	Prob > F	
Model	138.43	9	15.38	4.59	0.0285	significant
X1-SP	6.12	1	6.12	1.83	0.2184	
X2-PFA	3.12	1	3.12	0.93	0.3663	
X3-SF	0.50	1	0.50	0.15	0.7107	
X1 X2	4.00	1	4.00	1.19	0.3107	
X1 X3	2.25	1	2.25	0.67	0.4395	
X2 X3	2.25	1	2.25	0.67	0.4395	
$X1^2$	0.85	1	0.85	0.25	0.6294	
$X2^2$	53.06	1	53.06	15.84	0.0053	
$X3^2$	60.80	1	60.80	18.15	0.0037	
Residual	23.45	7	3.35			
Lack of Fit	18.25	3	6.08	4.68	0.0851	not significant
Pure Error	5.20	4	1.30			
Cor Total	161.88	16				

As shown in Table 7, the squares for SP, PFA, and SF were 6.12, 3.12, and 0.50, respectively, which shows a significant effect on the compressive strength of the single factors: SP has the greatest effect on the compressive strength of the concrete, PFA has the second greatest effect, and SF has the least effect. Similarly for the interaction, SP and PFA had the most significant effect on the compressive strength of the manufactured sand concrete, and the interaction between SP and SF and PFA and SF had the same effect on the compressive strength.

3.2. Prediction Model Validation

The reliability of the prediction model is verified through experimentation. Four sets of concrete blocks with different SP, PFA, and SF contents were created to measure the uniaxial compressive strength and were then compared with the prediction model. The content of SP, PFA, and SF used for the model validation experiments is shown in Table 8. The block-making process and compressive strength testing of the manufactured sand concrete blocks for the model validation experiments were the same as the methods previously mentioned in Section 2.2. The compressive strength experiments, predicted values, and relative errors are shown in Figure 5.

Table 8. The SP, PFA, and AF content of the validation experiments.

No.	SP	PFA	SF
1	7%	12%	2.5%
2	10%	15%	6%
3	12%	17%	2.5%
4	15%	20%	7%

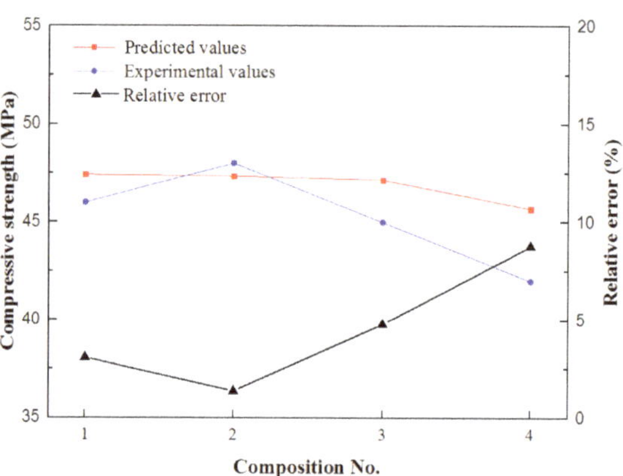

Figure 5. Compressive strength experiments, predicted values, and relative errors.

As can be seen from Figure 5, the relative error between the experimental and predicted values is less than 10%; therefore, the predictive model for the compressive strength of the manufactured sand concrete developed in this investigation can be considered to be credible.

3.3. Response Surface and Contour Analysis

The effect of the interaction between SP, PFA, and SF on the compressive strength was analyzed using response surface and contour analysis based on the regression equation for the compressive strength of the manufactured sand concrete. Another factor was controlled for at an intermediate level when discussing the pattern of interaction effects on compressive strength. The intermediate levels of SP, PFA, and SF content in this work were 10%, 15%, and 5% of the cement mass, respectively, as described in the previous section.

3.3.1. Effect of SP and PFA Interaction

Keeping the SF at an intermediate level, the response surface and contour in SP-PFA are shown in Figure 6. As can be seen in Figure 6a, the entire response surface takes on an "arch" shape when the SF is 5%. This indicates that the compressive strength tends to increase and then decrease as the SP and PFA content increases, with a maximum value existing. As can be seen from the contour lines in Figure 6b, the contour lines in the upper right corner of the picture are more densely distributed, which indicates that when the SP content and the PFA content are high, the change in PFA content has a greater effect on the fluctuation of the compressive strength. It is worth noting that the higher and lower contents described here correspond to the ranges set in this study. In the case of PFA, for example, the content in this work is between 10% and 20% by mass of cement, so this is the higher PFA content described to indicate a content close to, but not exceeding, 20%. The same rule is followed in the subsequent discussion.

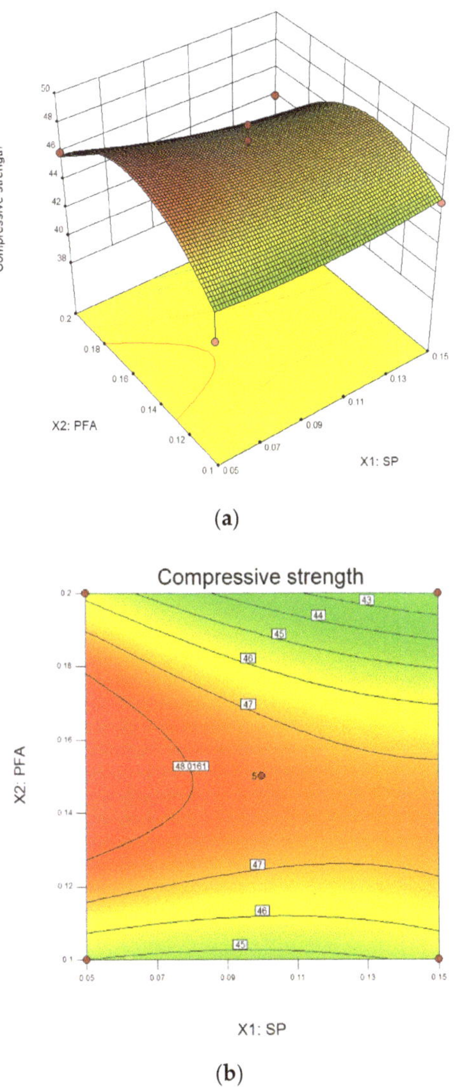

Figure 6. Response surface and contour in SP-PFA. (**a**) Response surface. (**b**) Contour.

3.3.2. Effect of SP and SF Interaction

Similarly, keeping the PFA at an intermediate level, the response surface and contour in SP-SF are shown in Figure 7. The response surfaces of Figures 6a and 7a both have an "arch" shape. Unlike Figures 6b and 7b has a higher degree of symmetry above and below the contour lines. In addition, as can be seen from the denseness of the contours in Figure 7b that the change in SF content has a greater effect on the fluctuation of the compressive strength when the SP content is high and the amount of SF is low. This means that when the PFA content is 15% of the cement mass and the SF content is less than 3.5% or greater than 6.5%, the effect of the change in SF content on the compressive strength fluctuates more than if the SF content is greater than 3.5–6.5%.

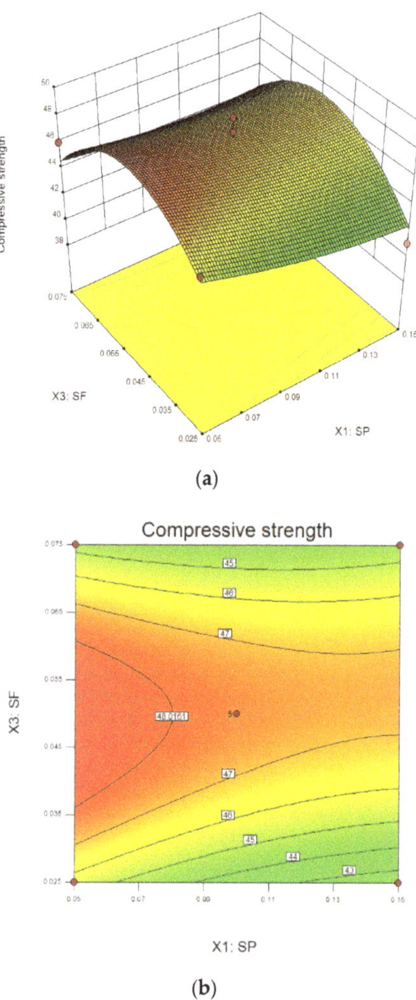

(**a**)

(**b**)

Figure 7. Response surface and contour in SP-SF. (**a**) Response surface. (**b**) Contour.

3.3.3. Effect of PFA and SF Interaction

When the SP is at an intermediate level, the response surface and contour in PFA-SF are shown in Figure 8. As can be seen from Figure 8a, the highest values exist across the response surface and are in the center, corresponding to a compressive strength of around 48 MPa for the manufactured sand concrete block. This indicates that in this study, the manufactured sand concrete exhibited a high compressive strength when the SP, PFA, and SF contents were all at intermediate levels. As can be seen in Figure 8b, the contours are centrosymmetric and equally spaced between the surrounding contours. This shows that equal variations in PFA and SF content have the same fluctuating effect on the compressive strength when the SP is at an intermediate level. In addition, the top leftmost corner of the contour in Figure 8b shows a compressive strength of less than 42 MPa, which is due to the high SF content and the low PFA content.

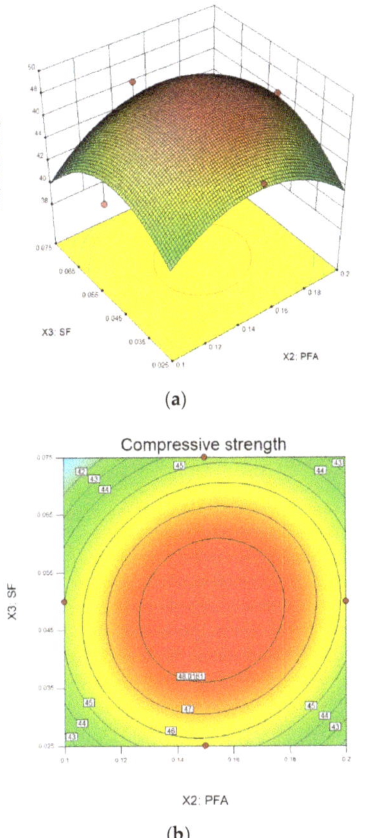

(**a**)

(**b**)

Figure 8. Response surface and contour in SFA-SF. (**a**) Response surface. (**b**) Contour.

4. Conclusions

An investigation of the compressive strength of manufactured sand concrete containing stone powder (SP), pulverized fuel ash (PFA), and silicon fume (SF) using the Box–Behnken experiment design in the response surface method (RSM) provides the following conclusions:

(1) A prediction model of the compressive strength of manufactured sand concrete with SP, PFA, and SF content was developed using multiple regression analysis with SP, PFA, and SF content as factors and compressive strength as the response value; the multiple correlation coefficient R^2 of the prediction model was 0.85. The prediction model for the compressive strength of concrete using manufactured sand was validated experimentally, and the validation results showed that the prediction model was credible, with a relative error of less than 10% between the experimental and predicted values.

(2) The statistical values of the single factors were analyzed and the degree of significance of the single factors on the compressive strength showed that SP content had the greatest effect on the compressive strength of the manufactured sand concrete, with PFA having the next greatest effect, and SF having the least effect. For the interactions, SP and PFA content had the most significant effect on the compressive strength of the manufactured sand concrete, while the interactions between SP and SF and PFA and SF had the same effect on the compressive strength.

(3) Response surface and contour analyses were carried out where the SP, PFA, and SF contents were kept at moderate levels (10%, 15%, and 5% of cement mass, respectively). The results show that the compressive strength tends to increase and then decrease with increasing SF and PFA content, with a maximum value. The maximum compressive strength of the concrete was found when the SP, PFA, and SF contents were all at intermediate levels.

Author Contributions: Conceptualization, H.M. and G.M.; methodology, H.M.; software, Z.S.; validation, H.M. and Z.S.; formal analysis, H.M.; investigation, H.M. and Z.S.; resources, Z.S. and H.M.; writing—original draft preparation, H.M.; writing—review and editing, H.M. and G.M.; supervision, G.M.; project administration, H.M., Z.S. and G.M. All authors have read and agreed to the published version of the manuscript.

Funding: This research was funded by Major scientific and technological innovation projects of Shandong Province (Grant No. 2019SDZY0304).

Institutional Review Board Statement: Not applicable.

Informed Consent Statement: Not applicable.

Data Availability Statement: The figures, tables and data that support the findings of this study are mentioned in the corresponding notes, with reference numbers and sources, and are publicly available in the repository.

Conflicts of Interest: The authors declare no conflict of interest.

Abbreviations

Designation	Explanation
RSM	response surface method
SP	stone powder
PFA	pulverized fuel ash
SF	silicon fume

References

1. Chen, L.; Sun, Z.; Liu, G.; Ma, G.; Liu, X. Spraying characteristics of mining wet shotcrete. *Constr. Build. Mater.* **2022**, *316*, 125888. [CrossRef]
2. Sun, Z.; Chen, L.; Yu, X.; Liu, G.; Pan, G.; Li, P.; Ma, H. Study on optimization of shotcrete loading technology and the diffusion law of intermittent dust generation. *J. Clean. Prod.* **2021**, *312*, 127765. [CrossRef]
3. Ly, H.-B.; Pham, B.T.; Dao, D.V.; Le, V.M.; Le, L.M.; Le, T.-T. Improvement of ANFIS Model for Prediction of Compressive Strength of Manufactured Sand Concrete. *Appl. Sci.* **2019**, *9*, 3841. [CrossRef]
4. Ji, T.; Chen, C.-Y.; Zhuang, Y.-Z.; Chen, J.-F. A mix proportion design method of manufactured sand concrete based on minimum paste theory. *Constr. Build. Mater.* **2013**, *44*, 422–426. [CrossRef]
5. Shen, W.; Yang, Z.; Cao, L.; Cao, L.; Liu, Y.; Yang, H.; Lu, Z.; Bai, J. Characterization of manufactured sand: Particle shape, surface texture and behavior in concrete. *Constr. Build. Mater.* **2016**, *114*, 595–601. [CrossRef]
6. Li, B.; Ke, G.; Zhou, M. Influence of manufactured sand characteristics on strength and abrasion resistance of pavement cement concrete. *Constr. Build. Mater.* **2011**, *25*, 3849–3853. [CrossRef]
7. Huang, Y.; Wang, L. Effect of Particle Shape of Limestone Manufactured Sand and Natural Sand on Concrete. *Procedia Eng.* **2017**, *210*, 87–92. [CrossRef]
8. Mundra, S.; Sindhi, P.R.; Chandwani, V.; Nagar, R.; Agrawal, V. Crushed rock sand—An economical and ecological alternative to natural sand to optimize concrete mix. *Perspect. Sci.* **2016**, *8*, 345–347. [CrossRef]
9. Wang, H.; Gao, M.; Gao, Y.; Chen, Q. Experimental study on dynamic characteristics of calcareous sand solidified by polymer. *J. Shandong Univ. Sci. Technol.* **2021**, *40*, 9.
10. Gonçalves, J.P.; Tavares, L.M.; Toledo Filho, R.D.; Fairbairn, E.M.R.; Cunha, E.R. Comparison of natural and manufactured fine aggregates in cement mortars. *Cem. Concr. Res.* **2007**, *37*, 924–932. [CrossRef]
11. Hong, H. Application of double mixing of mineral powder and fly ash in concrete. *Guangdong Archit.* **2010**, *2*, 3.
12. Luo, Z.; Zhang, X.; Wu, L.C. C60 super high-rise pumping concrete with admixture of key technology research. *Concrete* **2012**, *4*, 86–88.
13. Sun, Y.; Zhao, X.; Wang, S.; Chen, Z.; Liu, F.; Wu, L.; Li, C. Current Research Status on Application of Ultra-fine Recycled Building Materials Powder in Concrete. *J. Shandong Univ. Sci. Technol.* **2016**, *35*, 7.

14. Prakash Rao, D.S.; Giridhar Kumar, V. Investigations on concrete with stone crusher dust as fine aggregate. *Indian Concr. J.* **2004**, *78*, 45–50.
15. Li, B.; Zhou, M.; Cai, J.; Wang, S. Effect of microfines in manufactured sand on properties of various strength grade concretes. *Concrete* **2008**, *7*, 51–57.
16. Skaropoulou, A.; Kakali, G.; Tsivilis, S. Thaumasite form of sulfate attack in limestone cement concrete: The effect of cement composition, sand type and exposure temperature. *Constr. Build. Mater.* **2012**, *36*, 527–533. [CrossRef]
17. Schmidt, T.; Lothenbach, B.; Romer, M.; Neuenschwander, J.; Scrivener, K. Physical and microstructural aspects of sulfate attack on ordinary and limestone blended Portland cements. *Cem. Concr. Res.* **2009**, *39*, 1111–1121. [CrossRef]
18. Zhu, W.; Sun, Z. The affection to the performance of concrete cracking of fly ash and slag. *Fly Ash Compr. Util.* **2015**, *6*, 29–32.
19. Zhang, H.; Tan, X.; Ma, H.; Chen, S. Application of Mechanical-crushed Sands to Tunnel Shotcrete and Secondary Lining. *Tunn. Constr.* **2017**, *37*, 8.
20. Jain, N. Effect of nonpozzolanic and pozzolanic mineral admixtures on the hydration behavior of ordinary Portland cement. *Constr. Build. Mater.* **2012**, *27*, 39–44. [CrossRef]
21. Box, G.; Behnken, D.W. Some new three level designs for the study of quantitative variables. *Technometrics* **1960**, *2*, 455–475. [CrossRef]
22. Zhang, Q.; Feng, X.; Chen, X.; Lu, K. Mix design for recycled aggregate pervious concrete based on response surface methodology. *Constr. Build. Mater.* **2020**, *259*, 119776. [CrossRef]
23. Vázquez-Rivera, N.I.; Soto-Pérez, L.; St John, J.N.; Molina-Bas, O.I.; Hwang, S.S. Optimization of pervious concrete containing fly ash and iron oxide nanoparticles and its application for phosphorus removal. *Constr. Build. Mater.* **2015**, *93*, 22–28. [CrossRef]
24. Kumar, R. Modified mix design and statistical modelling of lightweight concrete with high volume micro fines waste additive via the Box-Behnken design approach. *Cem. Concr. Compos.* **2020**, *113*, 103706. [CrossRef]
25. Khudhair, M.H.; Al-Anweh, A.M.; Nomaan, M.H.; Berradi, M.; Elharfi, A. Response surface modeling of compressive strength of highperformance concrete formulated by a high water reducing and setting accelerating superplasticizer. Box-Behnken experimental design. *J. Chem. Technol. Metall.* **2019**, *54*, 135–144.
26. *GB/T 50081-2019*; Standard for Test Methods of Concrete Physical and Mechanical Properties. China Architecture and Building Press: Beijing, China, 2019.
27. Xue, D. Determination of uniaxial compressive strength of intact rock. *J. Shandong Univ. Sci. Technol.* **2020**, *39*, 9.

Article

Simulation of Two-Phase Flow of Shotcrete in a Bent Pipe Based on a CFD–DEM Coupling Model

Guanguo Ma [1,2], **Hui Ma** [2] and **Zhenjiao Sun** [1,*]

1   College of Safety and Environmental Engineering, Shandong University of Science and Technology, Qingdao 266590, China; ma156154682@126.com
2   College of Energy and Mining Engineering, Shandong University of Science and Technology, Qingdao 266590, China; hughie_ma@163.com
*   Correspondence: sunzhenjiao@foxmail.com

Abstract: To solve the problems in determining the interactions among particles and between particles and pipe walls in pneumatic conveying systems in field tests, this article studied the two-phase flow motion characteristics of shotcrete in pipes based on a CFD–DEM coupling model and field measurement. The movement of the shotcrete, which is affected by the gas phase in the pipe, was simulated for different bend angles, and the velocity of the shotcrete material and pressure distribution within the pipeline were determined. The simulation results show that at the ideal wind pressure, the inelastic collisions among the particles and between the particles and pipe wall cause the accumulation of shotcrete material in the outside area of the bent pipe section, which may block the pipe; nevertheless, the blockage is prevented by the turbulent and secondary flows, which disperse the particles to different degrees. In addition, the wear amounts caused by particles in pipes with different bend angles were quantified. With increasing bend angle, the wear points gradually diffuse radially toward the outside wall of the bent pipe. Additionally, the wear loss decreases and then increases with increasing bend angle. The particle velocity exhibits the minimal loss at a bend angle of 90°. It was concluded that the energy loss of the aggregate particles in the elbow of the pipe is approximately 30 times that in a horizontal, straight pipe. The results of this study can provide guidance in the construction field and for numerical simulations of the pneumatic conveying process of shotcrete.

Keywords: shotcrete materials; pneumatic conveying; CFD–DEM; pipe block; bend angle

Citation: Ma, G.; Ma, H.; Sun, Z. Simulation of Two-Phase Flow of Shotcrete in a Bent Pipe Based on a CFD–DEM Coupling Model. *Appl. Sci.* **2022**, *12*, 3530. https://doi.org/10.3390/app12073530

Academic Editor: N.C. Markatos

Received: 2 March 2022
Accepted: 29 March 2022
Published: 30 March 2022

Publisher's Note: MDPI stays neutral with regard to jurisdictional claims in published maps and institutional affiliations.

1. Introduction

Significant progress has been made in the theoretical and equipment research fields and the process of improvement and development of shotcrete applications. The wet-concrete spraying technology has been widely promoted and applied to above-ground and underground support structures [1–3]. Pumping is the most important pipeline transportation method for wet shotcrete in coal mines. The movement of the shotcrete material is characterized by the interactions among the particles and between the particles and pipe wall during pneumatic conveying in the rubber hose after pumping. In addition, determining the shotcrete material motion, material blockage, and wear of the bent pipe in the pipeline transportation process in a field test is difficult [4–6]. Thus, the pneumatic conveying process in the section of spraying concrete requires further investigation.

In the wet-shotcrete support project, the concrete material is usually pumped and pneumatically conveyed to the working face by a wet spraying machine (Figure 1). Pneumatic conveying is affected by many factors such as workability of concrete, environmental conditions, pipe assembly, actual pressure, and pump outlet speed; therefore, the field situation is more complex. Identifying the relevant factors, introducing them into numerical models, and evaluating their impacts on conveying performance are crucial and challenging. Therefore, the numerical simulations do not focus on the entire wet spraying system; instead, they are limited to the local area of the concrete flow, i.e., the piping units. The continuous spraying of underground concrete causes the bending deformation of the rubber hose. The impact of the shotcrete material on the bent pipe, their collision, and the shotcrete accumulation have adverse effects on the long-term use of the rubber hose and its spraying performance. Therefore, the movement characteristics of shotcrete material in the pipeline during pneumatic conveying must be studied.

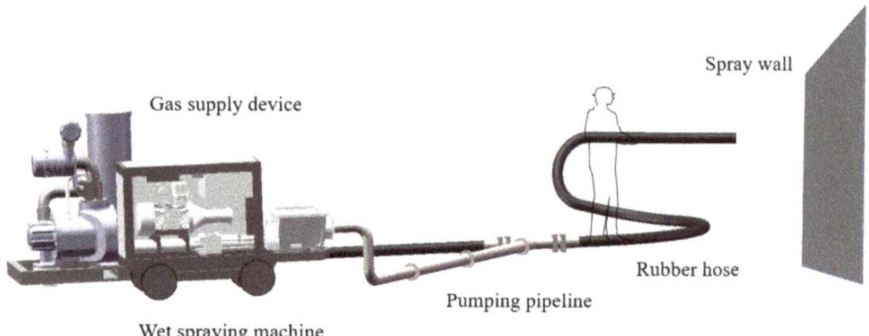

Figure 1. Setup for wet-shotcrete spraying.

Owing to the rapid advances in computer technology, numerical simulations have been widely used in the pneumatic conveying field. The discrete element method (DEM) is a numerical method that is used to simulate a great number of mutually interacting particles based on their equation of motion in a system. The DEM is usually coupled with computational fluid dynamics (CFD) in the study of gas–solid two-phase flows [7–12]. For example, Zhao [13] combined the CFD and DEM to simulate numerically the pneumatic conveying process in a horizontal channel and to study the effects of the wall roughness and random effects of turbulence on the particle diffusion. Du [14] used the CFD–DEM to simulate numerically the flow characteristics of gas–solid two-phase flows in a pneumatic conveying pipe with a bent section. Moreover, Oesterle [15] established a transport equation for the interface of gas–liquid coupled multiphase flow and used the numerical simulation software Fluent to simulate the velocity and pressure distributions at the interface between slug and gas–liquid two-phase flows.

Shotcrete is a mixture of cement slurry and has aggregates with a certain viscosity. Wu, Knut, and Li [16–18] used the DEM to model concrete materials with good accuracy. Moreover, Zhang [19] and Karakurt [20] established DEM models for self-compacting concrete in the filling state of the rock-fill structure and verified the model with experiments; in addition, they studied the workability and feasibility of the self-compacting concrete. Remond [21] developed a hard-core soft-shell model to simulate the concrete flow and determined the relationships among the main parameters of the mechanical model and the macroscopic rheological properties of the concrete numerical model; however, the concrete material was greatly simplified in the simulations. Hærvig [22] proposed a method for predicting the collisions of viscous particles based on the Hertz method to reduce the particle stiffness in the DEM simulations; this model improves the Johnson–Kendall–Roberts model (JKR model) and reduces the calculation cost; thus, the movement of a great number of viscous particles can be simulated. Ji [23,24] conducted a field test to study the pressure decrease and flow characteristics of solid–gas two-phase flow in a horizontal tube; the pressure decrease was measured with a pressure transmitter. In addition, they verified the test results with the coupled CFD–DEM and analyzed the effect of the pressure decrease on the bent section; however, the model was greatly simplified in the simulation, and the accelerated transportation stage of the conveyed material was not investigated.

Much effort has been devoted to investigating the flow characteristics of shotcrete in experiments [25–29]; nevertheless, the multiphase flow characteristics have not been thoroughly clarified. Because of the limiting experimental conditions, the microscopic mechanism of the particle–fluid interactions has not been fully understood. Although the coupled CFD–DEM has been applied in the modeling of pneumatic conveying processes, the pneumatic conveying process of shotcrete has rarely been studied, and simulations of the entire pneumatic conveying motion of shotcrete materials based on the CFD–DEM coupling model has not been reported.

In this study, the coupled CFD–DEM was used to simulate numerically the pneumatic conveying of shotcrete material in a rubber hose and to determine the characteristics of the gas phase flow field. Furthermore, the motion characteristics of the concrete material were studied, and the mechanical properties of the particles for different rubber hose structures were compared.

2. Governing Equations

In the pneumatic conveying process of wet-shotcrete materials, the movement state of wet-shotcrete materials and gas in the pipeline is extremely complex. To ensure the feasibility of simulation, shotcrete materials are usually discretized, as studied by Jiang, Chen [30–32], et al. Therefore, the process is simplified as gas–solid two-phase flow in this paper.

2.1. The Continuous Phase Model

The gas flow is treated as a continuous phase, the equation for conservation of mass, or continuity equation, can be written as follows

$$\nabla \cdot \left(\vec{v} \right) = S_m \tag{1}$$

$$\rho \left(\frac{\partial}{\partial t} \left(\vec{v} \right) + \nabla \cdot \left(\vec{v}\,\vec{v} \right) \right) = -\nabla p + \nabla \cdot (\tau) + \rho \vec{g} + \vec{F} \tag{2}$$

where S_m is the mass added to the continuous phase from the dispersed second phase, ρ is fluid density, $\vec{v}$ is fluid velocity, p is the static pressure, τ is the stress tensor (described below), and $\rho \vec{g}$ and $\vec{F}$ are the gravitational body force and external body forces, respectively. $\vec{F}$ also contains other model-dependent source terms such as porous-media and user defined sources.

In this paper, the transportation model selected was the standard $k - \varepsilon$ model. The turbulence kinetic energy, k and its rate of dissipation, ε are obtained from the following transport equations [33–35]:

$$\frac{\partial}{\partial t}(\rho k) + \frac{\partial}{\partial x_i}(\rho k u_i) = \frac{\partial}{\partial x_j} \left[\left(\mu + \frac{\mu_t}{\sigma_k} \right) \frac{\partial k}{\partial x_j} \right] + G_k + G_b - \rho \varepsilon - Y_M + S_k \tag{3}$$

$$\frac{\partial}{\partial t}(\rho \varepsilon) + \frac{\partial}{\partial x_i}(\rho \varepsilon u_i) = \frac{\partial}{\partial x_j} \left[\left(\mu + \frac{\mu_t}{\sigma_\varepsilon} \right) \frac{\partial \varepsilon}{\partial x_j} \right] + C_{1\varepsilon} \frac{\varepsilon}{k}(G_k + C_{3\varepsilon} G_b) - C_{2\varepsilon} \rho \frac{\varepsilon^2}{k} + S_\varepsilon \tag{4}$$

In these equations, G_k represents the generation of turbulence kinetic energy due to the mean velocity gradients, G_b is the generation of turbulence kinetic energy due to buoyancy, Y_M represents the contribution of the fluctuating dilatation in compressible turbulence to the overall dissipation rate, $C_{1\varepsilon}$, $C_{2\varepsilon}$, and $C_{3\varepsilon}$ are the turbulent Prandtl numbers for k and ε. S_k and S_ε are user-defined source terms.

2.2. Discrete Phase Model

The DEM was used to determine the particle phase of the shotcrete material. Based on the force that is acting on the shotcrete material particles, the particle velocity of the shotcrete material can be obtained with Newton's second law. The particle moves in the fluid by translational and rotational motions, and the force acting on the particle comprises several components. The equation of the equilibrium state of the shotcrete particles can be expressed as follows:

$$m_i \frac{dV_i}{dt} = F_{D,i} + F_{C,i} \tag{5}$$

$$I_i \frac{d\omega_i}{dt} = \sum_{i=1}^{n} T_i \tag{6}$$

where m_i is the mass of particle i, V_i the velocity of particle i, I_i the moment of inertia of particle i, ω_i the angular velocity of particle i, $F_{C,i}$ the collision force of particle i, $F_{D,i}$ the drag force acting on particle i, and T_i the torque acting on particle i.

2.3. CFD–DEM Coupling Model

In each time step, the DEM is used to calculate the position and velocity of each particle. After the iterative DEM calculation, the porosity and particle–fluid interaction force per unit volume in the fluid grid are calculated. The resulting data are used to calculate the fluid flow field and drag force of the fluid acting on the individual particles. Subsequently, the iterative CFD simulation is run until the value of the result converges; the result is applied to the discrete elements to obtain information about the particle motion in the next time step. The coupled, iterative CFD–DEM calculation process is shown in Figure 2.

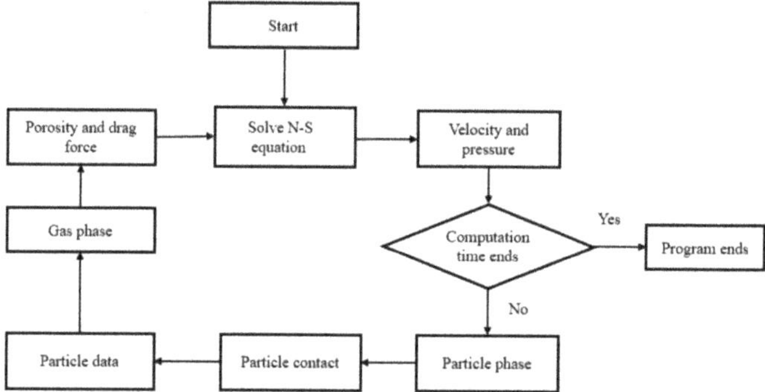

Figure 2. Flowchart of coupled, iterative CFD–DEM calculation.

The CFD–DEM gas–solid two-phase flow is embodied in the drag force between the gas–solid phases, and the gas–solid phase is coupled by the force acting between the two phases. In this study, the gas drag toward the particle phase and its reaction force were considered. Therefore, the commonly used Wen–Yu drag model was applied:

$$\beta_{\text{Wen-Yu}} = \frac{3}{4} C_D \frac{\rho_g \varepsilon_g \varepsilon_s |U_g - U_s|}{d_p} \varepsilon_g^{-2.65} \tag{7}$$

$$C_D = \begin{cases} \frac{24}{Re_p}\left(1 + 0.15 Re_p^{0.687}\right), & Re_p < 1000 \\ 0.44, & Re_p \geq 1000 \end{cases} \tag{8}$$

where Re_p is the Reynolds number of the particle ($Re_p = \rho_g \varepsilon_g \varepsilon_s |U_g - U_s| d_p / \mu_g$), d_p the particle diameter, C_D the drag coefficient, β the phase momentum exchange coefficient, ρ_g the fluid density, ε_g the fluid volume fraction, ε_s the solid volume fraction, U_g the fluid velocity, U_s the minimal spouting velocity, F the drag force acting on the particle, and μ_g the dynamic viscosity of the fluid.

2.4. Particle Contact Model

The JKR contact theory is the extension of the Hertz contact theory. It assumes that the adhesion effect only occurs on the contact surface. When two particles do not adhere to each other, their contact radius a_0 is expressed with the Hertz contact theory; otherwise, the external load is N, and the contact surface radius is a > a$_0$.

To determine the effect of the adhesion force on the contact properties, Johnson [36] applied the Griffith energy method to determine the surface energy $U_s = -\pi a^2 \Delta \gamma$, where the total energy U_T of the system is a function of the contact area A. When $\frac{dU_T}{dA} = 0$, U_T is in the equilibrium state. The equivalent load force N_1 of the two particles affected by the external load N and surface adhesion can be determined as follows:

$$N_1 = N + 3\pi R^* \Delta \gamma + \sqrt{(3\pi R^* \Delta \gamma)^2 + 6\pi R^* \Delta \gamma N} \tag{9}$$

The corresponding contact surface radius is as follows:

$$a^3 = \frac{3R^*}{4E^*}\left[N + 3\pi R^* \Delta \gamma + \sqrt{(3\pi R^* \Delta \gamma)^2 + 6\pi R^* \Delta \gamma N}\right] \tag{10}$$

The amount of overlap α of the two particles can be calculated with Equation (11):

$$\alpha = \frac{a^2}{R^*} - \left(\frac{2\pi a \Delta \gamma}{E^*}\right)^{1/2} \tag{11}$$

The increase in the normal force ΔN with respect to the increase in the amount of overlap $\Delta \alpha$ can be written as follows:

$$\Delta N = 2aE^* \Delta \alpha \left(\frac{3\sqrt{N} - 3\sqrt{N_c}}{3\sqrt{N} - \sqrt{N_c}}\right) \tag{12}$$

Based on Equation (12), there are two conditions:

For $\Delta \gamma = 0$, Equation (12) can be simplified to the Hertz contact force $a^3 = \frac{3R^*N}{4E^*}$ without considering the adhesion of the particles.

For $N = 0$, the two particles adhere to each other, and the radius of the contact surface can be expressed as follows:

$$a^3 = \frac{9\pi \Delta \gamma (R^*)^2}{2E^*} \tag{13}$$

When $(3\pi R^* \Delta \gamma)^2 + 6\pi R^* \Delta \gamma N \geq 0$, Equation (1) can be solved. The result is presented in Equation (14):

$$N \geq -\frac{3\pi R^* \Delta \gamma}{2} \tag{14}$$

Thus, when the external load N is negative (the two particles attract each other), the radius of the contact surface decreases; for $N \geq -\frac{3\pi R^* \Delta \gamma}{2}$, the particle adhesion is in a critical state, and when the pull force increases again, the two particles separate. The maximal pull force N_c required for separating the two particles is expressed as follows:

$$N_c = \frac{3\pi R^* \Delta \gamma}{2} \tag{15}$$

3. Geometric Model and Meshes

Rubber hoses provide flexibility to the pneumatic conveying process of concrete materials. However, bent pipe structures with different angles suffer from internal wear and unstable particle movement. The geometric model in Figure 3 consists of the horizontal pipes L_1 and L_2 and the bent-pipe structure R. The definite mass flow rate can be obtained based on the horizontal pipe L_1. The bend angles were set to $30°$, $45°$, $60°$, $90°$, $120°$, and $150°$. For the particle phase, the self-developed DEM code is merged into CFD, a coupling scheme is used between the CFD and DEM, and the explicit time integration method is used to calculate the motions of particles in the DEM. A compromise between efficiency and accuracy is made: spherical particles with diameters of 5 and 8 mm were used to represent the coarse aggregate covered with cement slurry. Based on field tests and related reports in the literature [37,38], the gas flow velocity was set to 64 m/s, and the initial particle velocity was 1.5m/s; subsequently, the particle was accelerated to a stable velocity for relatively stable flow. The applied particle parameters and operating conditions are listed in Table 1 [39–42].

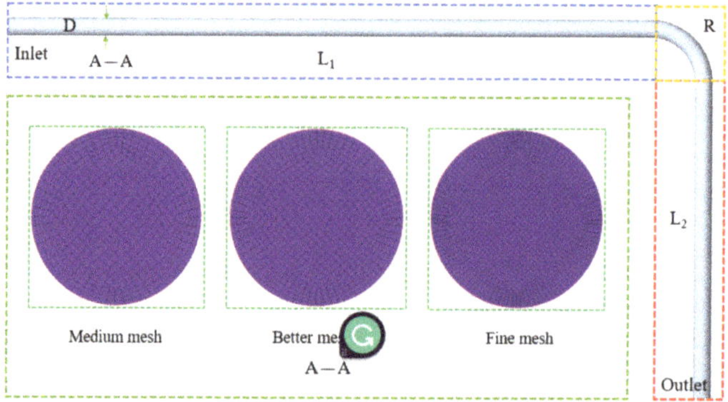

Figure 3. Geometric model and grids of a longitudinal section of bent pipe (bend angle of $90°$).

Table 1. Basic parameters.

Item	Index	Unit	Value
	Pipe diameter D	m	0.06
	Horizontal tube length L_1	m	5
	Horizontal tube length L_2	m	2.5
	Curvature radius of bent pipe R	m	0.2
	Coefficient of static friction	-	0.68
DEM	Coefficient of rolling friction	-	0.15
	Coefficient of restitution	-	0.55
	Density (particles)	kg/m^3	2300
	JKR surface energy	J	16
	Poisson ratio	-	0.25
	Time step	s	2×10^{-5}
	Density (air)	kg/m^3	1.225
	Viscosity	kg/(m·s)	1.7894×10^{-5}
CFD	Pressure outlet	Pa	101325
	Time step	s	2×10^{-3}
	Viscous model	-	k-epsilon model

The number of meshes has a great influence on the accuracy of the transient simulation results of the airflow field. Therefore, mesh independence studies are necessary to ensure that both discretization and rounding errors are within acceptable limits and that the mesh used does not significantly affect the simulation results. Three discrete grids with different densities are generated by ICEM, which are denoted as fine meshes, better meshes, and medium meshes, respectively. In all three cases, the mesh quality is above 0.5.

Using these three meshing cases as shown in Figure 3 (A—A section), the air velocities were compared and analyzed. Figure 4 shows the comparison results of velocity at cross-sections of the roadway. It can be observed that the simulation results of airflow velocity using three different kinds of grids show similar variation trends. The velocities using better meshes were very close to the values using fine meshes, but the results using medium meshes were quite different from the results using the other two different meshes. Therefore, it can be considered that better meshes have already met the requirement of meshing independence.

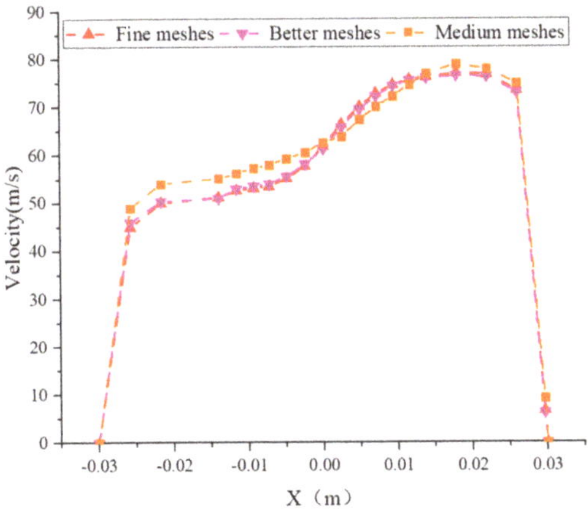

Figure 4. The comparison of the velocity among three meshing cases (bend angle of 90°).

4. Calculation Results and Analysis

4.1. Verification of Simulation

The gas power source adopts the air compressor with a rated power of 5.5 kW, maximum conveying air volume of 700 m^3/h, and rated pressure of 0.5 MPa to convey air. The inlet apparent gas velocity is controlled by the vortex flowmeter, which changes from 48 to 64 m/s. The field experiment is carried out with an SPB-7 wet concrete sprayer, and the pressure drop in the process of pneumatic conveying is measured by the differential pressure method. The layout diagram of the field test pressure transmitter is shown in Figure 5.

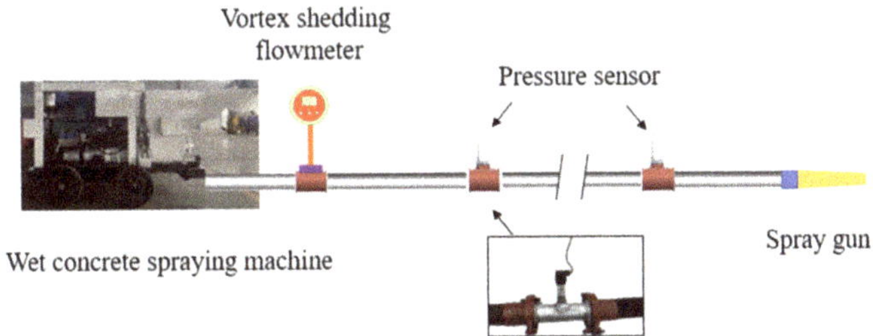

Figure 5. Layout diagram of the field test pressure transmitter.

The CFD–DEM coupling method can effectively observe the microscopic features that cannot be observed in the experiment, but the results must be consistent with the experimental results. This paper measured the pressure drop per unit length of the shotcrete in the 90° bent pipe at different air velocities when the concrete delivery volume was 4 m^3/h through field experiments, and simulated experiments using the same parameters, as shown in Figure 6.

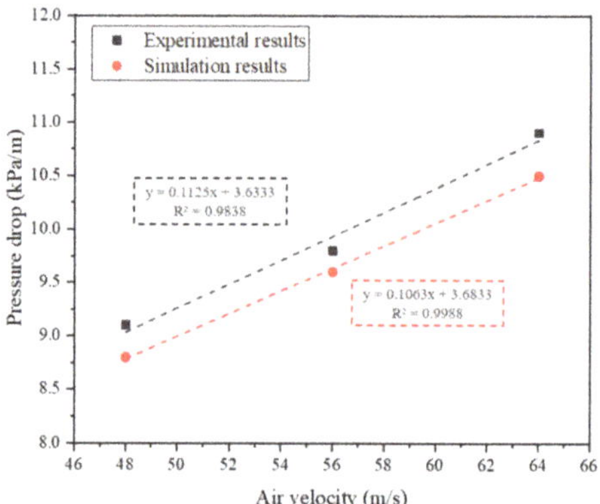

Figure 6. Variation in pressure drop with air velocity.

The simulation results were compared with the experimental results to verify the accuracy of the numerical method. The calculation results show that in the stable stage of pneumatic transportation of shotcrete, with the increase in airflow velocity, the pressure drop increases. The simulation results are in good agreement with the experimental results, the numerical values are consistent with the experimental values, and the errors are less than 10%. It shows that the model can be used for experimental calculation.

4.2. Gas Phase Flow Field

Figure 7 presents the global distribution of the gas phase static pressure during the pneumatic conveying process of the shotcrete material. The evident pressure loss within the rubber hose can be noted by comparing Figure 7, Figure 8a, and Figure 9. The pressure distribution in the horizontal, straight pipe section is relatively uniform compared to that near the bent pipe section. The pressures in the inside and outside areas of the bent pipe section are significantly different. The pressure in the outside area is high, and the pressure in the inside area is low, which is due to the centrifugal inertial force; in addition, the viscosity of the fluid on the pipe wall increases the fluid velocity on the wall surface. Moreover, both fluid velocities in the inside and outside areas are lower than that in the bent-pipe center, and the centrifugal inertial force of the fluid in the pipe center is greater than that in the inside and outside areas. The fluid in the pipe center flowing from inside to outside increases the external pressure, i.e., the pressure on the outside wall is much higher than that on the inside wall of the bent pipe. In addition, the gas velocity near the wall surface is relatively low, and the slow-flowing particles near the left and right boundaries move from the outside to the inside along the pressure decrease direction; the backflow at the central axis of the bent pipe results in a secondary flow. Subsequently, the secondary and main flows merge and move along the flow direction of the pipeline. The secondary flow limits the hitting velocity of the solid phase toward the outside wall of the bent pipe, and the central part of the outer surface of the bent pipe experiences more wear.

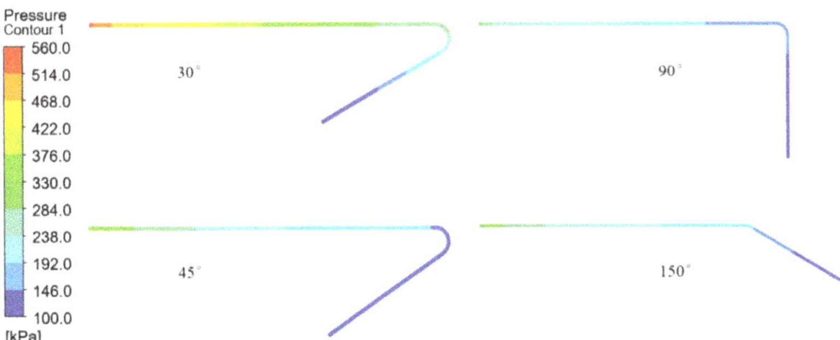

Figure 7. Global distribution of static pressure in pipelines with different bend angles.

According to Figure 8b, during the flow of the shotcrete material through the pipeline, the pressure in the straight pipe section in front of the bent pipe tends to decrease at a constant speed; behind the bent pipe section, the pressure decreases significantly for various bend angles. The pressure decrease in the 90° bent pipe is the lowest, whereas the other bent pipes show no evident pressure difference. During pneumatic conveying, the particle velocity within the pipeline varies significantly (Figure 9). Furthermore, the pressure difference affects the gas flow velocity distribution near the bent pipe, and the higher pressure pushes the gas phase to the side with the lower pressure. There are a higher pressure and a lower flow velocity in the outside area of the bent pipe part, and lower pressure and a higher flow velocity in the inside area of the bent pipe part.

4.3. Flow Field of Particle Phase

Figure 10 presents the particle velocity and particle distribution within the pipe under the action of a high-pressure gas flow. The particles within the horizontal pipe flow in a suspension state, and the particle packing density is not uniform. Under the action of a high-pressure gas flow, the particles have a higher density when they are just picked up by the gas flow. With the continuous generation of particles in the inlet, the pipe cross-section decreases, and the windward side of the particles increases. In addition, the packing density decreases continuously as the particles move forward. After entering the bent pipe, the particles accumulate near the midline of the outside area of the bent pipe; they form a rope-like structure (particularly in the 30° and 90° bent pipes), which is usually called "the particle rope" [43]; this typical phenomenon in gas–solid two-phase flows in bent pipes is mainly due to the inertia among particles and the inelastic collisions between the particles and wall surface. Under the actions of the subsequent turbulence and secondary flows, the particle rope starts to disperse gradually. The comparison of the particle flow characteristics in the three bent pipes shows that when passing through the 150° bent pipe, the particles disperse under the action of turbulence before forming the particle rope; this indicates the strong effect of the different angles of the bent pipes on the dispersion of the particle rope. Furthermore, after the particles pass through the bent pipe structure, their velocities recover at different times. The particles accumulate significantly later for a bend angle of 30° than for the other bend angles. Hence, the greater resistances among the particles and between the particles and pipe wall causes the particle velocities to recover more slowly. In addition, the particle accumulation at a bend angle of 30° is more severe and may block the pipe.

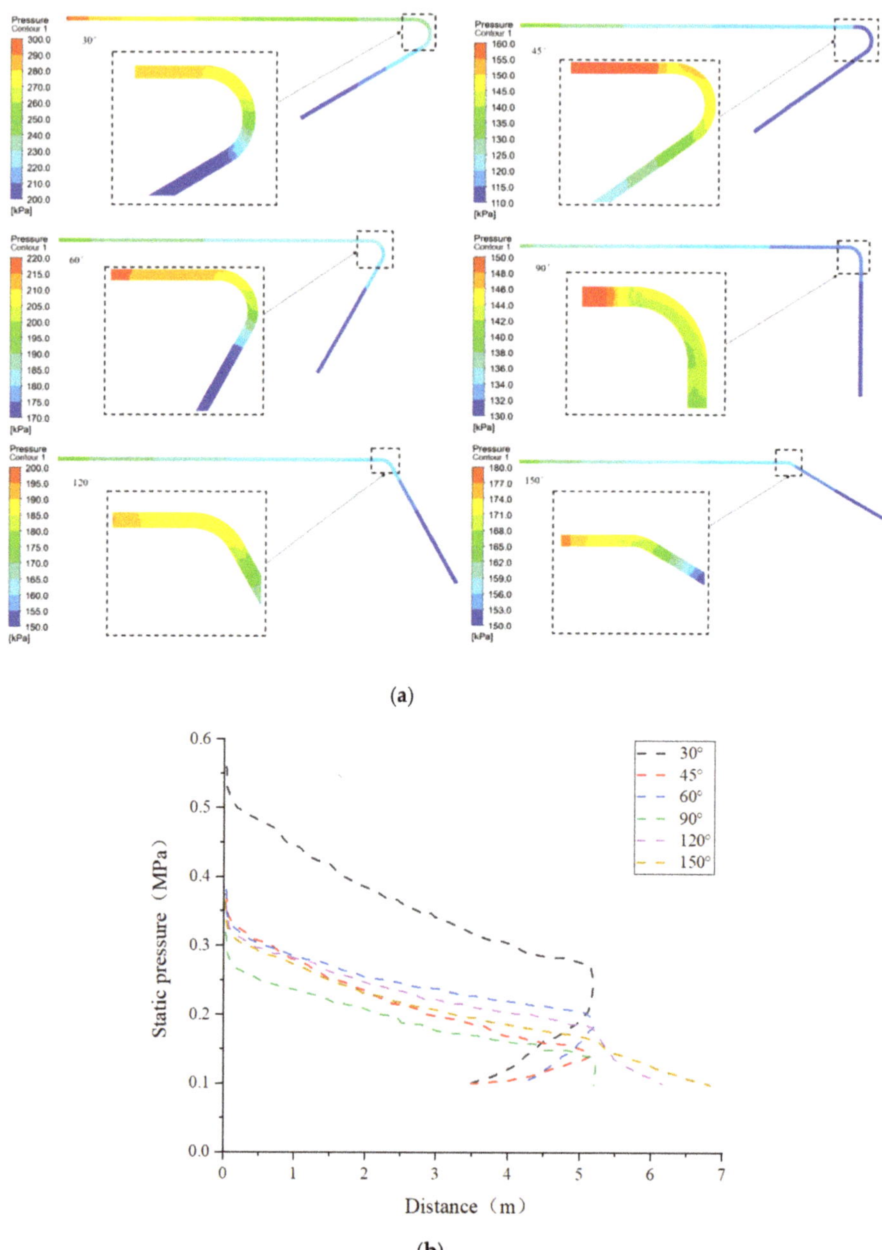

Figure 8. Distribution of static pressure in bent pipes with different bend angles. (**a**) z = 0 distribution cloud map of global static pressure in bent pipes with different bend angles. (**b**) Line chart of the global distribution of static pressure with respect to distance.

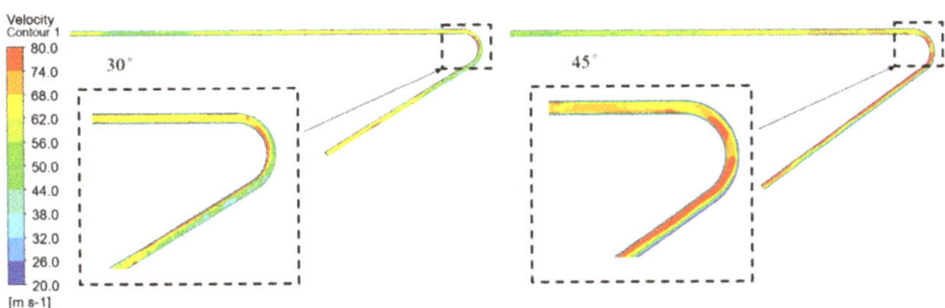

Figure 9. Distribution of gas phase velocities in 30° and 45° bent pipes.

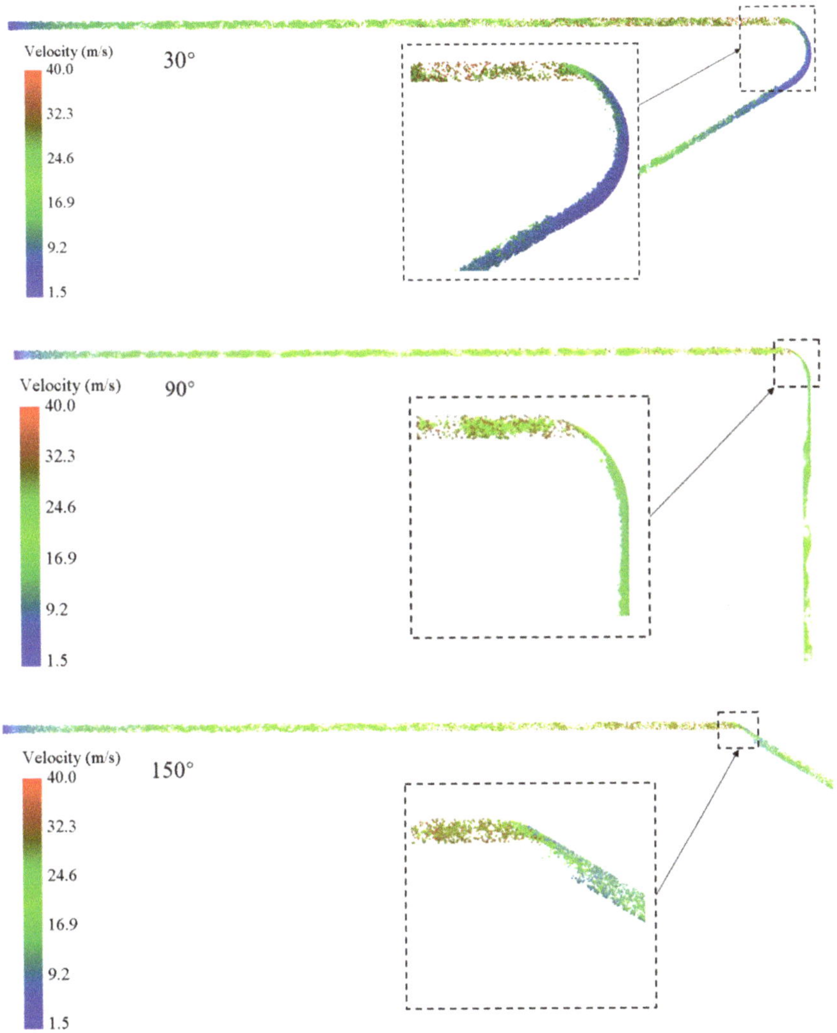

Figure 10. Distributions of moving concrete particles in 30°, 90°, and 150° bent pipes.

Figure 11a–c shows the variations in the velocities of individual concrete particles of different sizes in pipes with different bend angles over time. The particles comprise coarse aggregates with diameters of 5 and 8 mm (their surfaces are covered with cement slurry) and agglomerates of particles of different sizes. The values of 10 particles of each type were averaged to eliminate the accidental effect of particle sizes. According to Figure 11a–c, the particle velocity suddenly changes behind the bent pipe, and the 5 mm particles exhibit a greater acceleration than the other two particle types; hence, they enter the bent pipe section first. This is because although the particle with a large particle size has a larger area under the action of air, it can obtain more acceleration, but the resistance of air to particles is relatively high. Obviously, the resistance has a greater impact on the particle at an air velocity of 64 m/s.

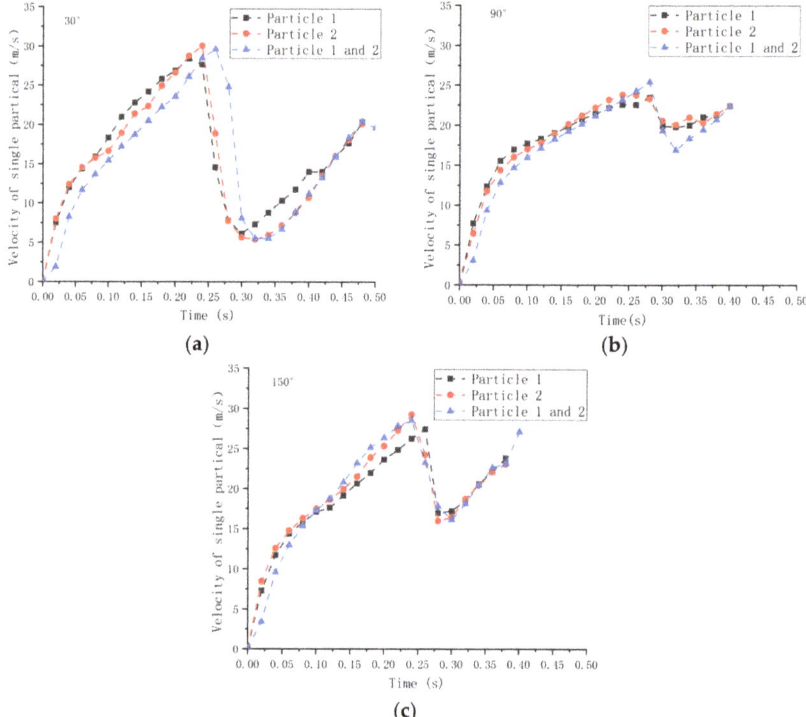

Figure 11. Variation in velocity of a single particle concerning time (Particle 1 represents the coarse aggregate with 8 mm diameter, Particle 2 represents coarse aggregate with 5 mm diameter, and Particles 1 and 2 represent agglomerates with two diameters). (**a**) Bend angle of 30°. (**b**) Bend angle of 90°. (**c**) Bend angle of 150°.

The different bent pipe structures have great effects on particle velocity. The particle velocity decreases by approximately 73% behind the 30° bent pipe and by approximately 15–35% behind the 90° and 150° bent pipes. This is because the particle in the 30° bent pipe is not affected by the gas phase, and its velocity continues to decrease. When the particle leaves the bent pipe section, it is picked up again by the airflow and accelerated under the actions of turbulence and secondary flows. When the particle passes through the 90° bent pipe, its movement is not heavily affected by the turbulence and secondary flows; thus, the particle velocity does not increase further and stabilizes. In addition, the turbulence and secondary flows in the bent pipe act as a resistance to the particles. Because the 90° bent pipe provides the lowest resistance, the particle velocity experiences the lowest loss.

The CFD–DEM model provides detailed data about particle movement and collisions. Figure 12 presents the main wear areas and wear effect of the concrete particles on the bent pipe with different bend angles. The results show that within the same time interval, the 45° bent pipe experiences significant wear, and its maximal depth of wear is 0.042 mm. In addition, the main wear position diffuses radially toward the outside wall of the bent pipe with an increasing bend angle. This is because the wall roughness promotes particle–wall collisions. The reflection angle at a rough wall is much greater than that at a smooth wall. The rougher wall causes more particles to diffuse radially along the outside wall of the lifted bent pipe, while the higher axial velocity causes the particles to suspend uniformly. Thus, when the particles hit the bent pipe, they are uniformly distributed across the curved section of the bent pipe, thereby causing elliptical wear; these results are consistent with those presented by Solnordal [44].

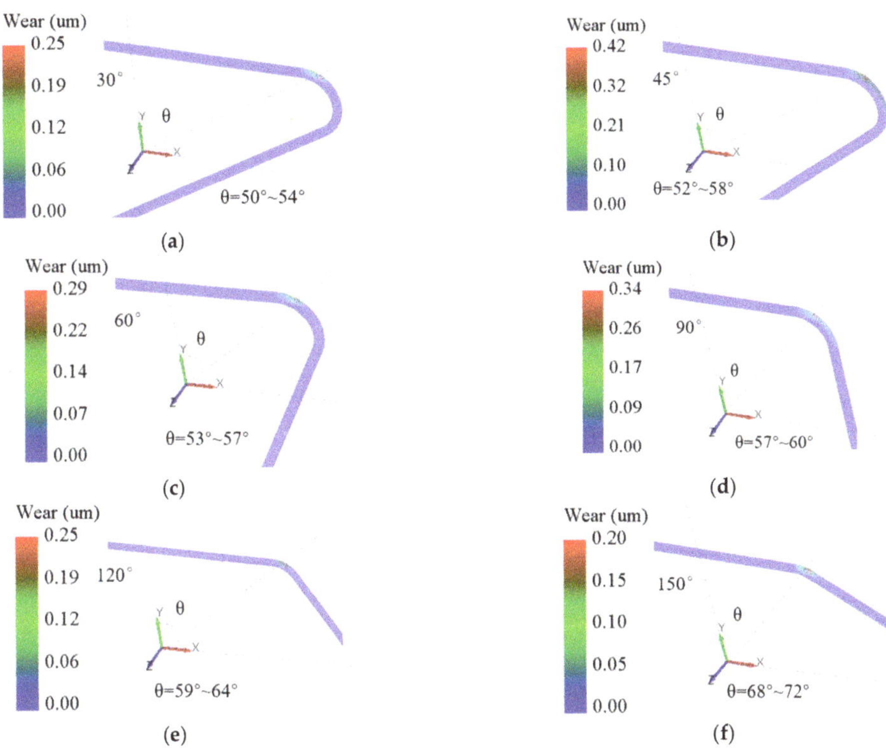

Figure 12. (a–f)Amount of wear and main wear position of particles in bent pipes with different angles.

Figure 13 shows the numbers of particle–pipe wall contact times and normal contact forces of the shotcrete particles in the pipes with different bend angles. With increasing bend angle, the number of particles–pipe wall contact times and normal contact force first increase and then decrease. At a bend angle of 45°, both values exhibit a maximum, which is consistent with the results of Xu [45]. In addition, in pneumatic conveying, the normal contact forces of most particles affect the bent pipe; it is assumed that the 45° bent pipe is subjected to stronger impact and wear.

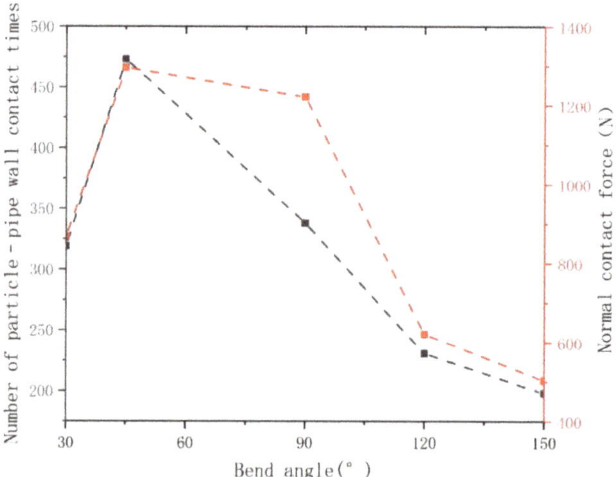

Figure 13. Number of particle–pipe wall contact times in the pipe in 0.5 s.

Figure 14 presents the variation in the particle energy loss concerning time. When the particles enter the bent pipe, the particle–pipe wall collisions significantly increase the energy loss. As the particle rope gradually forms, the number of particles–pipe wall contact times decreases, and the energy loss gradually decreases and stabilizes. The particle energy loss is not correlated with the bent pipe structure, and the particle energy loss in the bent pipe is approximately 30 times that in the straight pipe.

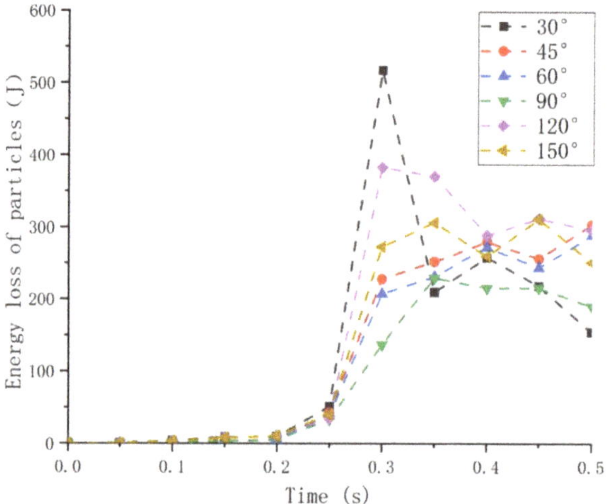

Figure 14. Variation in particle energy loss concerning time.

5. Conclusions

In this paper, the flow characteristics of wet-shotcrete materials in the horizontal hose were simulated, and a CFD–DEM model suitable for simulating the flow characteristics of wet-shotcrete materials was developed to study the flow of wet-shotcrete materials in the horizontal hose. In addition, the accuracy of the numerical simulation results was verified through the on-site pressure drop experimental test. The main conclusions are as follows:

(1) The pipeline exhibits a significant pressure loss and pressure difference at the inside and outside walls of the bent pipe. The pressure gradient force leads to a secondary flow, which limits the hitting velocity of the solid phase toward the outside wall of the bent pipe. In addition, the central part of the outer surface of the bent pipe experiences more wear.

(2) The bending angles have a great effect on the particle velocity. The particle loss decreases first and then increases with increasing bend angle. Owing to the inertia and inelastic collisions among the particles and between the particles and pipe wall, the concrete material particles accumulate in the outside area of the bent pipe section in the pneumatic conveying process, which may block the pipe. Nevertheless, the turbulence and secondary flows cause the aggregated particle rope to disperse gradually. The particle velocity decreases by more than 73%, and the particle is accelerated behind the 30° bent pipe section owing to the turbulence and secondary flows. When the bend angle is 90°, the turbulence and secondary flows have less effect on the particle movement behind the bent pipe, and the particle loss is minimal.

(3) With increasing bend angle, the number of particles–pipe wall contact times and the normal contact force firstly increase and then decrease; both reach a maximum at a bend angle of 45°. The wear impact on the 45° bent pipe is stronger than those on the other bent pipes, and the main wear position diffuses radially toward the outside wall of the bent pipe with increasing bend angle. Furthermore, the particle energy loss in the bent pipe is approximately 30 times that in the straight pipe.

Author Contributions: Investigation, H.M.; Methodology, G.M.; Resources, Z.S.; Validation, H.M.; Visualization, G.M.; Writing—original draft, G.M. and Z.S.; Writing—review & editing, Z.S. All authors have read and agreed to the published version of the manuscript.

Funding: This study was funded by projects such as Major scientific and technological innovation projects of Shandong Province (Grant No. 2019SDZY0203).

Institutional Review Board Statement: Not applicable.

Informed Consent Statement: Not applicable.

Data Availability Statement: Not applicable.

Conflicts of Interest: The authors declare that they have no conflict of interest.

References

1. Guoming, L.; Weimin, C.; Zhen, L.; Xin, Y. Pressure Distribution Calculation Along Pipes Used for Mining Pumping Wet Shotcrete Based on Shear and Slip. *J. Shandong Univ. Sci. Technol. Nat. Sci.* **2017**, *6*, 66–72.
2. Yin, L.; Yalan, G.; Hao, L.; Haoyu, W. Experimental study on effect of recycled aggregate from construction waste on conveying performance of mine filling paste. *J. Shandong Univ. Sci. Technol.* **2020**, *3*, 59–65.
3. Qunfei, F.; Jie, F.; Jie, S.; Xiaoqiang, C.; Xianjun, L. Effect of respirable particulate matter on dust removal performance of polyacrylonitrile fiber bundle filter. *J. Shandong Univ. Sci. Technol.* **2019**, *6*, 33–39.
4. Tripathi, N.M.; Santo, N.; Kalman, H.; Levy, A. Experimental analysis of particle velocity and acceleration in vertical dilute phase pneumatic conveying. *Powder Technol.* **2018**, *330*, 239–251. [CrossRef]
5. Ghafori, H.; Sharifi, M. Numerical and experimental study of an innovative design of elbow in the pipe line of a pneumatic conveying system. *Powder Technol.* **2018**, *331*, 171–178. [CrossRef]
6. Chen, L.; Sun, Z.; Liu, G.; Ma, G.; Liu, X. Spraying characteristics of mining wet shotcrete. *Constr. Build. Mater.* **2022**, *316*, 125888. [CrossRef]
7. Roussel, N.; Geiker, M.R.; Dufour, F.; Thrane, L.; Szabo, P. Computational modeling of concrete flow: General overview. *Cem. Concr. Res.* **2007**, *37*, 1298–1307. [CrossRef]
8. Zhou, H.; Yang, Y.; Wang, L. Numerical investigation of gas-particle flow in the primary air pipe of a low NOx swirl burner—The DEM-CFD method. *Particuology* **2015**, *19*, 133–140. [CrossRef]
9. Cheikh, K.E.; Remond, S.; Khalil, N.; Aouad, G. Numerical and experimental studies of aggregate blocking in mortar extrusion. *Constr. Build. Mater.* **2017**, *145*, 452–463. [CrossRef]
10. Nerella, V.N.; Mechtcherine, V. Virtual Sliding Pipe Rheometer for estimating pumpability of concrete. *Constr. Build. Mater.* **2018**, *170*, 366–377. [CrossRef]

11. Ebrahimi, M.; Siegmann, E.; Prieling, D.; Glasser, B.J.; Khinast, J. An investigation of the hydrodynamic similarity of single-spout fluidized beds using CFD-DEM simulations. *Adv. Powder Technol.* **2017**, *28*, 2465–2481. [CrossRef]

12. Pietsch, S.; Heinrich, S.; Karpinski, K.; Muller, M.; Schonherr, M.; Jager, F.K. CFD-DEM modeling of a three-dimensional prismatic spouted bed. *Powder Technol.* **2017**, *316*, 245–255. [CrossRef]

13. Zhao, H.; Zhao, Y. CFD–DEM simulation of pneumatic conveying in a horizontal channel. *Int. J. Multiph. Flow* **2019**, *118*, 64–74. [CrossRef]

14. Du, J.; Hu, G.; Fang, Z.; Fan, Z. Simulation of dilute pneumatic conveying with bends by CFD-DEM. *J. Natl. Univ. Def. Technol.* **2014**, *4*, 134–139. [CrossRef]

15. Wang, X.; Sun, X. Three-dimensional simulations of air–water bubbly flows. *Int. J. Multiph. Flow* **2010**, *36*, 882–890. [CrossRef]

16. Wu, J.; Naggar, M.H.E.; Li, X.; Wen, H. DEM analysis of geobag wall system filled with recycled concrete aggregate. *Constr. Build. Mater.* **2020**, *238*, 117684. [CrossRef]

17. Krenzer, K.; Mechtcherine, V.; Palzer, U. Simulating mixing processes of fresh concrete using the discrete element method (DEM) under consideration of water addition and changes in moisture distribution. *Cem. Concr. Res.* **2019**, *115*, 274–282. [CrossRef]

18. Li, Z.; Cao, G.; Guo, K. Numerical method for thixotropic behavior of fresh concrete. *Constr. Build. Mater.* **2018**, *187*, 931–941. [CrossRef]

19. Zhang, X.; Zhang, Z.; Li, Z.; Li, Y.; Sun, T. Filling capacity analysis of self-compacting concrete in rock-filled concrete based on DEM. *Constr. Build. Mater.* **2020**, *233*, 117321. [CrossRef]

20. Karakurt, C.; Celik, A.O.; Yilmazer, C.; Kiricci, V.; Ozyasar, E. CFD simulations of self-compacting concrete with discrete phase modeling. *Constr. Build. Mater.* **2018**, *186*, 20–30. [CrossRef]

21. Remond, S.; Pizette, P.J.C.; Research, C. A DEM hard-core soft-shell model for the simulation of concrete flow. *Cem. Concr. Res.* **2014**, *58*, 169–178. [CrossRef]

22. Haervig, J.; Kleinhans, U.; Wieland, C.; Spliethoff, H.; Jensen, A.L.; Sorensen, K.; Condra, T.J. On the Adhesive JKR Contact and Rolling Models for Reduced Particle Stiffness Discrete Element Simulations. *Powder Technol.* **2017**, *319*, 472–482. [CrossRef]

23. Ji, Y.; Liu, S.; Li, J. Experimental study of supply pressure on spraying materials in horizontal pipe. *Vacuum* **2019**, *159*, 51–58. [CrossRef]

24. Ji, Y.; Liu, S. Effect of secondary flow on gas-solid flow regimes in lifting elbows. *Powder Technol.* **2019**, *352*, 397–412. [CrossRef]

25. Liuabc, G.; Chengab, W.; Chenab, L.; Panab, G.; Liuab, Z. Rheological properties of fresh concrete and its application on shotcrete. *Constr. Build. Mater.* **2020**, *243*, 118180.

26. Pan, G.; Li, P.; Chen, L.; Liu, G. A study of the effect of rheological properties of fresh concrete on shotcrete-rebound based on different additive components. *Constr. Build. Mater.* **2019**, *224*, 1069–1080. [CrossRef]

27. Chenab, L.; Zhanga, X.; Liua, G. Analysis of dynamic mechanical properties of sprayed fiber-reinforced concrete based on the energy conversion principle. *Constr. Build. Mater.* **2020**, *254*, 119167. [CrossRef]

28. Cui, X.; Liu, G.; Wang, C.; Qi, Y. Effects of PET Fibers on Pumpability, Shootability, and Mechanical Properties of Wet-Mix Shotcrete. *Adv. Civ. Eng.* **2019**, *2019*, 2756489. [CrossRef]

29. Chen, L.; Zhou, Z.; Liu, G.; Cui, X.; Dong, Q.; Cao, H. Effects of substrate materials and liner thickness on the adhesive strength of the novel thin spray-on liner. *Adv. Mech. Eng.* **2020**, *12*, 1687814020904574. [CrossRef]

30. Jiang, S.; Chen, X.; Cao, G.; Tan, Y.; Xiao, X.; Zhou, Y.; Liu, S.; Tong, Z.; Wu, Y. Optimization of fresh concrete pumping pressure loss with CFD-DEM approach. *Constr. Build. Mater.* **2021**, *276*, 122204. [CrossRef]

31. Chen, L.; Sun, Z.; Ma, H.; Pan, G.; Li, P.; Gao, K. Flow characteristics of pneumatic conveying of stiff shotcrete based on CFD-DEM method. *Powder Technol.* **2022**, *397*, 117109. [CrossRef]

32. Ma, G.; Sun, Z.; Ma, H.; Li, P. Calibration of Contact Parameters for Moist Bulk of Shotcrete Based on EDEM. *Adv. Mater. Sci. Eng.* **2022**, *2022*, 6072303. [CrossRef]

33. Sun, Z.; Chen, L.; Yu, X.; Liu, G.; Pan, G.; Li, P.; Ma, H. Study on optimization of shotcrete loading technology and the diffusion law of intermittent dust generation. *J. Clean. Prod.* **2021**, *312*, 127765. [CrossRef]

34. Yan, F.; Luo, C.; Zhu, R.; Wang, Z. Experimental and numerical study of a horizontal-vertical gas-solid two-phase system with self-excited oscillatory flow. *Adv. Powder Technol.* **2019**, *30*, 843–853. [CrossRef]

35. Wei, J.; Zhang, H.; Wang, Y.; Wen, Z.; Yao, B.; Dong, J. The gas-solid flow characteristics of cyclones. *Powder Technol.* **2017**, *308*, 178–192. [CrossRef]

36. Johnson, K.L.; Kendall, K.; Roberts, A. Surface Energy and the Contact of Elastic Solids. *Proc. R. Soc. A Math. Phys. Eng. Sci.* **1971**, *324*, 301–313.

37. Chen, L.; Ma, G.; Liu, G.; Liu, Z. Effect of pumping and spraying processes on the rheological properties and air content of wet-mix shotcrete with various admixtures. *Constr. Build. Mater.* **2019**, *225*, 311–323. [CrossRef]

38. Liu, G.; Guo, X.; Cheng, W.; Chen, L.; Cui, X. Investigating the migration law of aggregates during concrete flowing in pipe. *Constr. Build. Mater.* **2020**, *251*, 119065. [CrossRef]

39. Cross, R. Measurements of the horizontal coefficient of restitution for a superball and a tennis ball. *Am. J. Phys.* **2002**, *70*, 482–489. [CrossRef]

40. Li, T.; Zhang, J.; Ge, W. Simple measurement of restitution coefficient of irregular particles. *China Particuology* **2004**, *2*, 274–275. [CrossRef]

41. Grima, A.P.; Wypych, P.W. Investigation into calibration of discrete element model parameters for scale-up and validation of particle-structure interactions under impact conditions. *Powder Technol.* **2011**, *212*, 198–209. [CrossRef]
42. Barrios, G.K.P.; De Carvalho, R.M.; Kwade, A.; Tavares, L.M. Contact parameter estimation for DEM simulation of iron ore pellet handling. *Powder Technol.* **2013**, *248*, 84–93. [CrossRef]
43. Chu, K.; Yu, A. Numerical Simulation of the Gas−Solid Flow in Three-Dimensional Pneumatic Conveying Bends. *Ind. Eng. Chem. Res.* **2008**, *47*, 7058–7071. [CrossRef]
44. Solnordal, C.B.; Wong, C.Y.; Boulanger, J. An experimental and numerical analysis of erosion caused by sand pneumatically conveyed through a standard pipe elbow. *Wear* **2015**, *336*, 43–57. [CrossRef]
45. Xu, L.; Zhang, Q.; Zheng, J.; Zhao, Y. Numerical prediction of erosion in elbow based on CFD-DEM simulation. *Powder Technol.* **2016**, *302*, 236–246. [CrossRef]

applied sciences

Article

Time–Frequency Domain Characteristics of Acoustic Emission Signals and Critical Fracture Precursor Signals in the Deep Granite Deformation Process

Le Zhang [1,2], Hongguang Ji [1,2,*], Liyuan Liu [1,2,*] and Jiwei Zhao [1,2]

1   College of Civil and Resource Engineering, University of Science and Technology Beijing, Beijing 100083, China; le_changl@163.com (L.Z.); 17801203090@163.com (J.Z.)
2   Beijing Key Laboratory of Urban Underground Space Engineering, University of Science and Technology Beijing, Beijing 100083, China
*   Correspondence: jihongguang@ces.ustb.edu.cn (H.J.); liuliyuan@ustb.edu.cn (L.L.)

Featured Application: This article describes the frequency-domain characteristics of acoustic emission signals of granite in ultra-deep gold mine.

Abstract: To study the crack evolution law and failure precursory characteristics of deep granite rocks in the process of deformation and failure under high confining pressure, granite samples obtained from a depth of 1150 m are tested using a TAW-2000 triaxial hydraulic servo testing machine and a PCI-II acoustic emission monitoring system. Based on the stress–strain curve and IET function, the loading process of the sample is divided into five stages: crack closure, linear elastic deformation, microcrack generation and development, macroscopic fracture generation and energy surge, and post-peak failure. The evolution trend and fracture evolution law of the acoustic emission signal event interval function in different stages are analyzed. In particular, the signals with an amplitude greater than 85 dB, a peak frequency greater than 350 kHz, and a frequency centroid greater than 275 kHz are defined as the failure precursor signals before the rock reaches the peak stress. The defined precursor signal conditions agree well with the experimental results. The time–frequency analysis and wavelet packet decomposition of the precursor signal are performed on the extracted characteristic signal of the failure precursor. The results show that the time-domain signal is in the form of a continuous waveform, and the frequency-domain waveform has multi-peak coexistence that is mainly concentrated in the high-frequency region. The energy distribution obtained by the wavelet packet decomposition of the characteristic signal is verified with the frequency-domain waveform. The energy distribution of the signal is mainly concentrated in the 343.75–375 kHz frequency band, followed by the 281.25–312.5 kHz frequency band. The energy proportion of the high-frequency signal increases with the confining pressure.

Keywords: rock mechanics; deep granite; acoustic emission; precursory characteristics; time–frequency analysis

Citation: Zhang, L.; Ji, H.; Liu, L.; Zhao, J. Time–Frequency Domain Characteristics of Acoustic Emission Signals and Critical Fracture Precursor Signals in the Deep Granite Deformation Process. *Appl. Sci.* **2021**, *11*, 8236. https://doi.org/10.3390/app11178236

Academic Editors: Xiangming Hu and Guoming Liu

Received: 14 July 2021
Accepted: 31 August 2021
Published: 5 September 2021

Publisher's Note: MDPI stays neutral with regard to jurisdictional claims in published maps and institutional affiliations.

1. Introduction

The rapid developments of science and technology, the increasing energy demand, and hundreds of years of exploitation are presently leading to the gradual exhaustion of energy resources on the earth's surface [1]. Theoretically, the available metallogenic space in the earth's interior is distributed from the surface to 10,000 m below the surface. To cater to the requirements of social development, deep mining on the earth is a strategic scientific and technological matter that calls for research focus [2]. The deep rock exists in an environment of high temperature, geostress, and pore water pressure, and its microstructure and macroscopic properties largely differ from those of shallow rock [3]. In the deep mining process, specific damage phenomena such as rockburst and spalling may occur [4,5]. To ensure the safety and stability of deep rock mass in the mining process, investigating the development law of crack initiation, propagation, and failure in deep rocks during loading is beneficial to understand the change law of deep rocks and obtain the rock failure precursor information in actual construction [6].

Solid materials such as rocks store strain energy in their internal structures. When rocks are subjected to external loads, their internal strain energy will be released rapidly in the form of elastic waves, referred to as acoustic emission (AE) [7–9]. AE information analysis methods in the rock fracture process mainly include parameter analysis and waveform analysis. Parameter analysis is the most commonly used AE signal analysis method [10–17]. Typical AE parameters include event number, absolute energy, duration, and rise time. In addition to these basic parameters, b-value [18–20], average frequency (AF) value (AE counts/duration), and RA value (rise time/amplitude) [21–23] can also represent the failure mode and damage process of rocks or concrete materials. However, parameter analysis only provides a simple description of waveform characteristics. By comparison, waveform analysis is an in-depth analysis based on the time–frequency domain of the signal and explains the failure mechanism and precursors from the waveform spectrum characteristics of the signal. In recent years, waveform analysis has been studied extensively. Xu used Fourier transform to analyze microseismical signals and concluded that the rockburst signal is the superposition of low-frequency and high-frequency signals [24]. D.g.aggelis discussed the acoustic emission characteristics of different fracture modes and concluded that tensile failure signals prefer higher frequency and shorter waveform than shear failure signals [25]. This study analyzed the acoustic emission data of sedimentary limestone and found that a large number of high-amplitude and low-frequency events occurred near the failure of the rock sample [26]. Li studied the two characteristic frequency bands of the AE waveform obtained from a marble tensile test [27]. Zhang used the peak frequency of AE signals as a parameter to classify them by fuzzy C-means and studied the distribution of different signals in the loading process feature [28]. Wang studied the AE frequency-domain characteristics of granite under freezing–thawing cycles and revealed that the rock structure degrades when subjected to high frequency–temperature cycles, showing the characteristics of relatively high number of low-frequency signals and small intervals [29].

Most of the studies on AE characteristics during rock deformation and failure focus on the time- and frequency-domain parameter variability of AE signals from shallow rocks. Therefore, to analyze the mechanical properties and failure precursors of deep granite under high confining pressure, this study uses the time–frequency evolution of acoustic emission signals in the triaxial test of granite obtained from 1150 m depth underground in Sanshandao Gold Mine, Laizhou, Shandong Province, China. Based on the time–frequency characteristics of AE signals and mechanical tests, the AE signal distribution and the deformation and failure process of deep granite under different confining pressures are studied. Furthermore, the failure precursor signal before peak stress is defined. The research results provide some theoretical guidance for rock failure prediction in the actual construction of acoustic emission technology.

2. Test Process and Design

A triaxial compression test was conducted on granite samples to study the AE information, precursor characteristics, and energy distribution of the samples during loading deformation and failure under high confining pressure. The granite samples were all excavated from 1150 m underground in Sanshandao Gold Mine in Laizhou, Shandong, China. All samples were collected from the same rock mass, with uniform mass and without obvious joint layers, thereby avoiding the effect of varied samples on the test results. According to the method recommended by ISRM, the sample was prepared into a cylinder with a diameter of 50 mm and a height of 100 mm, maintaining the unevenness error of both ends of the sample less than 0.05 mm and the diameter error along the sample height less than 0.3 mm. In general, when the deformation ratio of the testing machine is inconsistent with that of the granite sample in the loading process, the friction at the sample end will limit the end deformation of the sample, thus affecting the mechanical properties, such as the stress distribution of the sample. To eliminate this sample end effect, we polished both ends of the sample with 400 grit emery paper, reducing the friction of the end faces. The in situ stress measured at −1150 m is 30 MPa. To simulate the real rock occurrence environment and conduct the control test, the confining pressure in this test was preset to 10, 20, and 30 MPa. The size of each granite sample was measured and recorded as shown in Table 1.

Table 1. Mechanical parameters of deep granites.

Specimen	Height/mm	Diameter/mm	Weight/g	Density/g/cm^3	Confining Pressures/MPa
A1	100.81	50.04	518	2.613	
A2	100.53	50.05	509	2.575	
A3	100.28	50.21	514	2.590	10
A4	100.96	50.14	523	2.625	
B1	100.30	50.03	516	2.617	
B2	100.21	50.31	512	2.572	
B3	100.46	50.07	521	2.635	20
B4	100.75	50.02	516	2.608	
C1	100.08	50.19	525	2.653	
C2	100.65	50.00	513	2.597	
C3	100.12	50.22	517	2.608	30
C4	100.36	50.06	515	2.609	

The test equipment includes a rock mechanics loading system and an AE system (Figure 1). The loading system adopts a TAW-2000 triaxial hydraulic servo testing machine. During the loading process, the preset confining pressure value was first applied, and then the axial pressure on the rock sample was gradually increased until its failure under the constant confining pressure. In the later stage of loading, to obtain a clear post-peak curve, the loading rate was reduced at 0.008 mm/s until the sample was destroyed. For AE signal acquisition, the PCI-II acoustic emission monitoring instrument and its software were used. Vaseline was smeared on the end of the AE monitoring probe to ensure the coupling effect between the probe and the sample. The sampling evaluation rate of the AE monitoring system was set at 1 MSPS, the acquisition threshold was 40 dB, and the front gain was 40 dB to eliminate the effect of external noise.

(a)

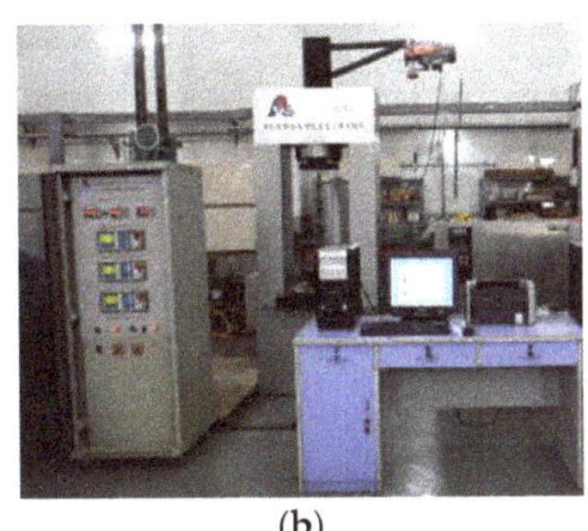
(b)

(c)

Figure 1. Rock mechanics test system and acoustic emission test layout. (**a**) PCI-II acoustic emission system; (**b**) TAW-2000 triaxial hydraulic servo testing machine; (**c**) the layout of acoustic emission monitoring probe.

3. Experimental Phenomena and Mechanism Analysis

The peak strength, peak strain, and elastic modulus of granite samples A1 to C4 were obtained under confining pressures of 10, 20, and 30 MPa applied until sample failure. The samples under the same confining pressure show similar mechanical properties; hence, in this paper, we mainly analyze samples A1, B1, and C1. Figure 2 shows the stress–strain curves and cross-sections of these samples.

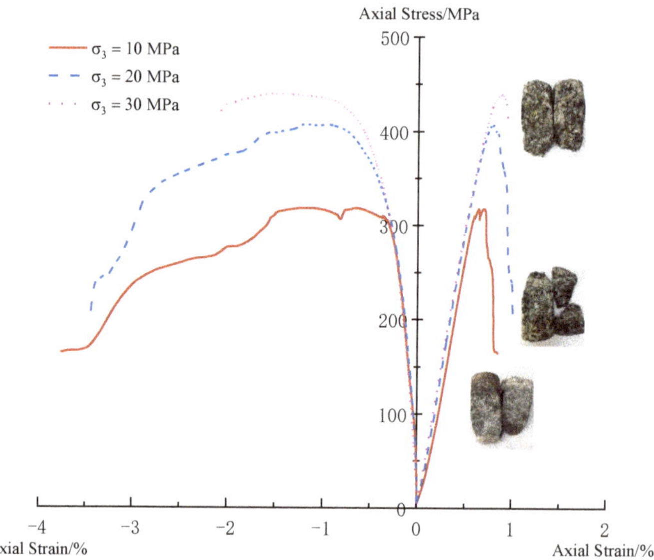

Figure 2. Stress–strain curve obtained from the triaxial test.

As shown in Figure 2, the deep granite samples mainly show elastic deformation until the peak stress, after which their strength decreases rapidly, indicating strong brittleness under different confining pressures. These rocks originate in an environment of high geostress, temperature, and hydraulic pressure, and under the action of low confining pressure, the main fracture surfaces of the rock samples exhibit obvious scratch marks, and many rock fragments are observed. Moreover, the opening degree of the main fracture is relatively large, which shows obvious signs of tension and distortion. This is mainly due to the secondary shear failure caused by the concentration of slip stress on the fracture surface after failure. As the confining pressure increases, the main fracture surface develops from the end to the sample's side. Under high confining pressure, the fracture surface of the rock sample is rough, which indicates a volume expansion phenomenon of the micro drum.

The fracture morphology of the three rock samples under triaxial compression was observed by scanning electron microscopy (SEM).

As shown in Figure 3, the granite samples' fracture morphology is extremely complex, with their surfaces showing diverse morphology consisting of defects. The whole rock mass comprises layers, which consist of defects between them. Further, rock debris accumulates on the defects. The fracture boundary is locally concentrated due to brittle fracture. The fracture of granite is characterized by cleavage fracture, which is mainly fluvial cleavage, with obvious scratches and slip marks. The fault is predominantly composed of intracrystalline cracks (with rock debris accumulation on the surface) and intergranular cracks.

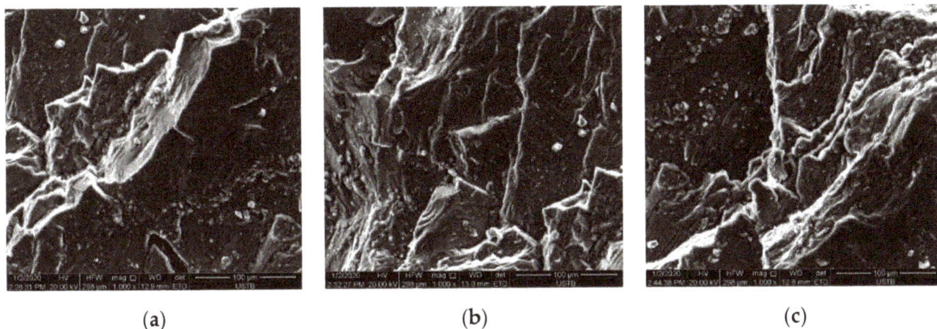

(a)   (b)   (c)

Figure 3. SEM scanning image. (**a**) A1, (**b**) B1, (**c**) C1.

Many primary fractures of different sizes exist in granite. When subjected to splitting load, the microcracks begin to sprout at small-sized cracks and hard components. With increasing load, these microcracks and main cracks begin to expand and connect into main cracks. The cracks propagate completely and produce an obvious fracture surface until the rock is destroyed. Therefore, the fracture surface of granite has various shapes and obvious surface cracks. Thus, cracks formed through cracks of varied size will produce a large amount of rock debris on the surface.

4. Analysis of Micro-Scale Failure Evolution of Deep Granite Based on IET Function

A large number of microcracks are randomly distributed in natural granite. When an external load is applied to the granite sample, the primary microcracks cause local stress concentration inside the sample, which further promotes the occurrence and propagation of internal microcracks. The expansion and aggregation of microcracks lead to the production of macroscopic cracks and local failure surfaces until the failure produces a large number of AE signals. These signals contain the evolution information of internal failure and fracture in the process of rock deformation and failure [20,30].

As all the samples were obtained from the same core, their stress–strain curves indicate that the axial stress–strain relationship and the AE event curve of all samples under different confining pressures shows similar trends. This reflects that the distribution of microcracks in the deep granite samples is basically similar, with no remarkable difference between the samples.

According to the deformation characteristics, the loading process of rock can be classified into five stages: (1) crack closure, (2) linear elastic deformation, (3) crack initiation and stable crack growth, (4) critical energy release and unstable crack growth, and (5) failure and post-peak behavior [31,32]. During rock loading, the AE event rate of each fracture level differs considerably. The AE event monitoring method has been widely used to classify and verify the degree of rock cracking [33,34]. To further explore the evolutionary relationship between the AE event number and the fracture degree, this study adopts the IET function $F(\tau)$ proposed by Triantis and Kourkolis [35,36].

The calculation method is as follows:

(1) Define a sliding time window with N impact event intervals

$$\left(t_{i-1}, t_{i+N-1}\right) \tag{1}$$

(2) Calculate the moving average of N event intervals in the specified time window:

$$\tau_i = \frac{t_{i+N-1} - t_{i-1}}{N}(i = 2, 3, \ldots) \tag{2}$$

special:

$$\tau_1 = \frac{t_N - t_1}{N}(i = 1) \tag{3}$$

(3) For a given time window:

$$F(\tau) = \tau_i^{-1}(i = 1, 2, 3, \ldots) \tag{4}$$

In this test analysis, we adopted $n = 50$, as explained by Triantis and Kourkolis. Here, N represents the resolution of AE characteristics and does not majorly affect the analysis results [36].

Figure 4 shows the stress–strain–$F(\tau)$ evolution curves for the three granite samples.

For sample A1, the value of the event interval function is extremely low and the curve is close to the horizontal axis within 0–2087 s, implying less AE activity. According to Z. Moradian [31], a small amount of AE activity is mainly caused by the compaction of primary pores or shear slip on the primary failure surface.

Within 784–1869 s, the stress curve of the rock sample is approximately a straight line, indicating that the rock is in the elastic loading stage. The microcracks close and destroy the cementation points on the rough grain surface of some fracture surfaces, and the $F(\tau)$ value increases slightly.

Within 1869–2087 s, the stress curve of the rock sample still shows a straight-line form, but with slightly decreasing slope. This is because the loading rate is reduced to 0.008 mm/s to obtain a clear post-peak curve in the loading stage. At the same time, the later value of $F(\tau)$ in this stage increases significantly, which indicates that the rock sample undergoes the formation and development of microcracks and the phenomenon of microcrack nucleation. This is the fracture generation and development stage.

Within 2087–3455 s, the stress curve no longer shows linear development, the stress value gradually increases to the maximum value, and a series of U-shaped evolution paths of the function $F(\tau)$ appears. The U-shaped evolution path of $F(\tau)$ is the path in which $F(\tau)$ decreases, then stays at a relatively low level, and finally increases with time. In this time period, the sample shows stable crack propagation, and the minority of macroscopic crack extension is limited because of the expansion of the macroscopic crack generation. The expansion of the macroscopic cracks of the frontier forms the surface for new strain energy dissipation. This strain energy is used to maintain the new microcrack generation at a high speed, which subsequently leads to the reduction of $F(\tau)$. After a series of cracking events occur, the value of $F(\tau)$ returns to a high level.

After 3455 s, the rock sample moves to the post-peak failure stage, and $F(\tau)$ initially remains at a high level, showing a typical unstable crack growth form. The function value $F(\tau)$ also shows a continuous and compact U-shaped evolution trend. The continuous occurrence of large $F(\tau)$ values indicates rapid increase in the AE activity simultaneously. Further, the accumulated energy of the granite sample begins to release during the loading process, and the internal cracks of the sample are connected under the action of sustained external force. This allows the rock sample to slide along the internal main fracture surface of the sample, resulting in local shear tensile failure and, finally, in the macroscopic failure of the sample.

According to the AE signal value $F(\tau)$ and the stress–strain curve of granite under compression, the loading process of the granite sample is divided into five stages: (1) crack closure, (2) linear elastic deformation, (3) microcrack generation and development, (4) macroscopic fracture generation and energy surge, and (5) post-peak failure. In stages 1 and 2, the rock sample experiences the compaction of primary pores, shear slip on the primary failure surface, relative slip between grains, and destruction of the rough particle surface at the microscale. After the end of the elastic loading stage, the stress–strain curve of the sample is no longer linear; the stress rapidly reaches the maximum value and a large number of microcrack propagation aggregates produce macroscopic cracks. In the post-peak stage, all three samples show high brittleness. In this stage, the large crack is quickly connected and formed. With continuously increasing loading, the main fracture surface inside the sample slides, causing local damage to the granite interior and complete damage eventually.

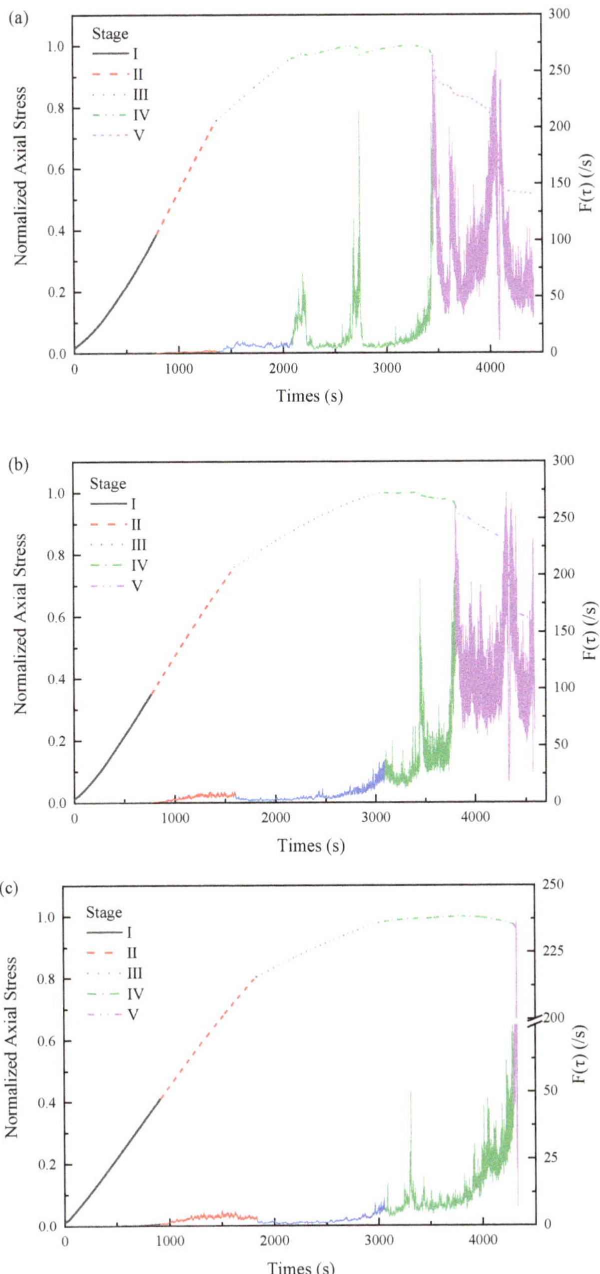

Figure 4. Stress–strain $F(\tau)$ evolution curves of granites. (**a**) A1, (**b**) B1, (**c**) C1.

5. Research on the Frequency-Domain Characteristics of AE Signals

The inhomogeneity and defects in the rock cause stress concentration in the loading process. Internal microcracks cause and expand or intensify plastic deformation, resulting in the rapid release of strain energy stored in the rock and subsequent AE. In practice, thousands of AE events occur when a sample is loaded until it breaks. Thus, the experimental and practical construction efficiency could be greatly improved by extracting characteristic signals from many AE signals that explain different degrees of rock fracture and determining the AE frequency-domain precursors of rock fracture.

5.1. Discussion on AE Signal Distribution and Peak Stress Precursor

The peak frequency of the AE signal is a characteristic parameter of the signal in the frequency domain. It is defined as the peak value observed in the power spectrum. The frequency centroid is also a frequency feature calculated, in kilohertz, as the sum of magnitude times frequency divided by the sum of magnitude, as equivalent to the first moment of inertia. As shown in Figure 5, the amplitude represents the maximum amplitude of the signal waveform, which is usually expressed in dB.

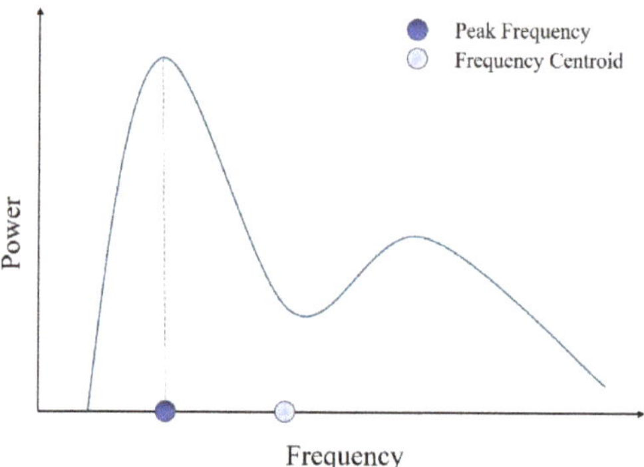

Figure 5. Configuration of peak frequency and frequency centroid.

In addition to using AE parameters to predict fracture precursors, this section studies the failure mechanism and precursory characteristics of critical failure by statistical analysis of the distribution of peak frequency, frequency centroid, and amplitude in the deformation and failure process of granite samples.

5.1.1. Frequency-Domain Distribution of AE Signals and the Definition of Prediction Signals

To study the variation of crack degree and peak AE signal frequency by the triaxial compression test under different confining pressures of the deep granite samples A1, B1, and C1, the relation scatter plots of amplitude, peak frequency, and frequency centroid with respect to normalized axial stress ($\sigma_1 / \sigma_{\max}$) were drawn, as shown in Figures 6–8, respectively, to acquire the frequency domain distribution law of deep granite AE signals and the precursory characteristics before reaching peak stress.

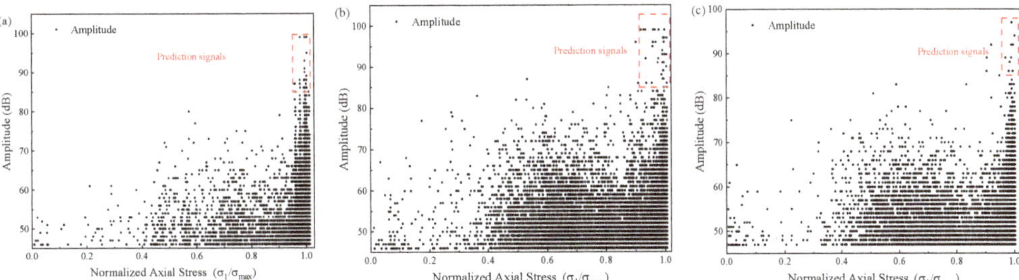

Figure 6. Relationship between amplitude distribution and normalized axial stress. (**a**) A1, (**b**) B1, (**c**) C1.

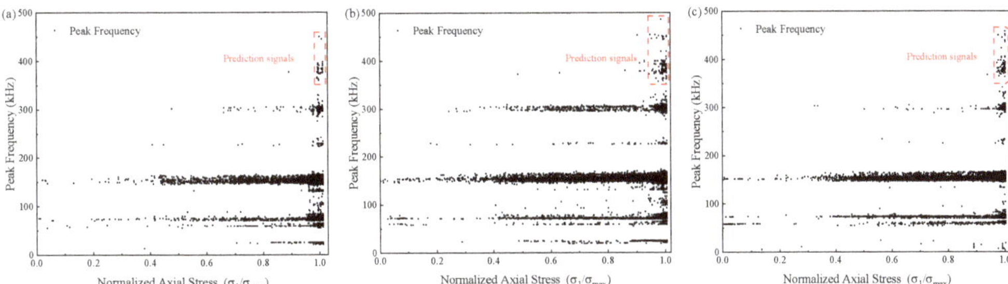

Figure 7. Relationship between peak frequency distribution and normalized axial stress. (**a**) A1, (**b**) B1, (**c**) C1.

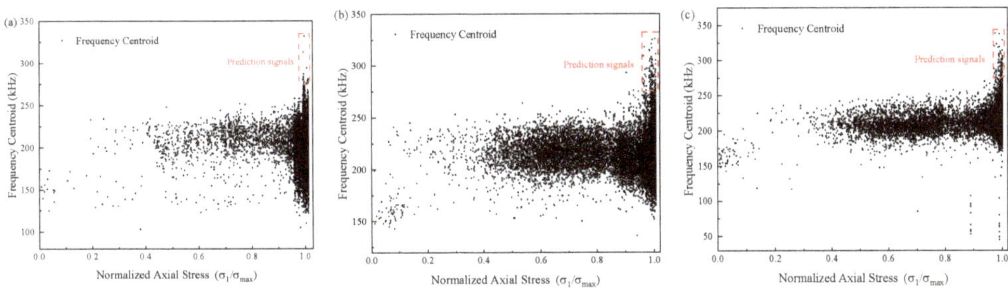

Figure 8. Relationship between frequency centroid distribution and normalized axial stress. (**a**) A1, (**b**) B1, (**c**) C1.

The figures indicate that during the test, the AE signals with an amplitude of 35–80 dB account for the majority. The peak frequency presents an obvious stratification phenomenon, while the distribution of frequency centroid band shows the law from thin to wide. In general, the distribution characteristics of the three AE parameters show some similar regularity during the test. Therefore, the study of the frequency-domain characteristics of AE signals before loading to the peak stress in the triaxial test can more comprehensively grasp the change patterns of fracture corresponding to various AE signals. It can better define the AE signals of rock failure prediction before loading to the peak stress from the perspective of the frequency domain.

The amplitude distribution of AE signals clearly shows that the number of AE signals is less in the crack closure stage and the early stage of linear elastic deformation. The corresponding results can also be obtained in the distribution of the other two signals. The maximum amplitude in each stage shows a similar trend with stress; it increases with increasing stress, before reaching the peak stress. Beyond this, the amplitude distribution breaks in the range of 35–80 dB and some signals larger than 80 dB begin to appear.

The peak frequency spectrum can be divided into three parts: low-frequency band (0–200 kHz), medium-frequency band (200–350 kHz), and high-frequency band (350–500 kHz). The low-frequency signals exist in the entire testing process. The frequency band around 75 and 160 kHz begins to widen with stress. When the normalized axial stress value reaches 50–60%, a large number of medium-frequency signals begin to appear and are concentrated around 225 and 300 kHz. Compared with the low- and medium-frequency signals, the practical significance of high-frequency band signals is more obvious. These signals appear almost at the same time when the amplitude crosses 80 dB, and only a little earlier than the time the stress peak is reached. Therefore, the appearance of signals with a peak frequency greater than 350 kHz has a strong correlation with rock failure.

As shown in Figure 5 and the concept of frequency centroid mentioned above, the higher the energy of AE signals in a certain frequency band, the closer the frequency centroid value will be to that frequency band. The frequency centroid distribution diagram shows that the AE energy is mainly concentrated in the frequency band of 150–175 kHz at the beginning of crack closure. With increasing stress, the AE energy is concentrated at approximately 225 kHz until the peak stress is reached. The frequency centroid shows the same characteristics as the values of amplitude and peak frequency. That is, its bandwidth widens, and most importantly, more signals larger than 275 kHz appear.

Hence, according to the above results, the failure precursor signal can be considered to be the signal with an AE amplitude greater than 85 dB, a peak frequency greater than 350 kHz, and a frequency centroid greater than 275 kHz; hereinafter, this signal is referred to as the P-signal. Comparison of the stress value corresponding to the P-signals with the peak stress of the granite samples reveals that the stress corresponding to the P-signals under the three confining pressures is 97%, 96%, and 98% of the peak stress values of the granite samples, respectively, and they occur 212, 370, and 393 s ahead of the occurrence of the peak stress in the test.

5.1.2. Time–Frequency Domain of the P-Signals

In Section 5.1.1, P-signals, with large peak frequency, frequency centroid, and amplitude, are defined as the precursor signals for rock instability and failure. Then, the P-signals in each confining pressure test were extracted and filtered, and the detailed frequency-domain information of the P-signals was obtained according to the fast Fourier transform (FFT). The time and frequency domain waveforms of the P-signals under different confining pressures are shown in Figures 9 and 10, respectively. The P-signals under each confining pressure show extremely similar characteristics; that is, the primary dominant frequency of the P-signal is concentrated near 375 kHz, while the secondary dominant frequency is mostly concentrated at 310 kHz. The frequency-domain signal shows the phenomenon of "multi-peak coexistence," dominated by the primary frequency.

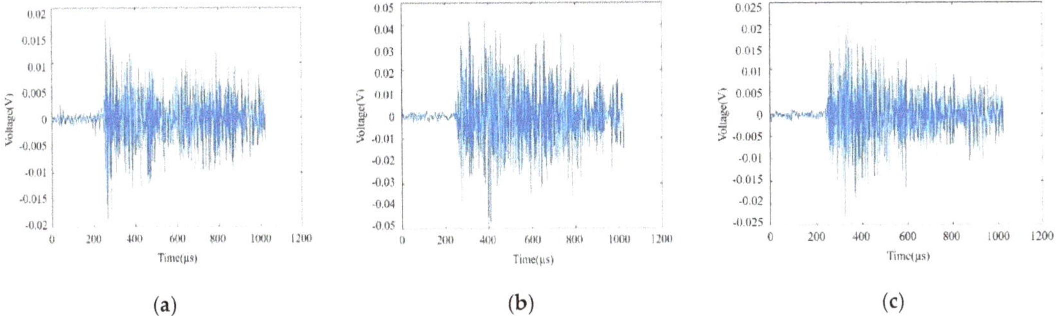

(a) (b) (c)

Figure 9. Time domain waveform of the P-signals of each sample. (**a**) A1.No.17715, (**b**) B1.No.8200, (**c**) C1.No.6375.

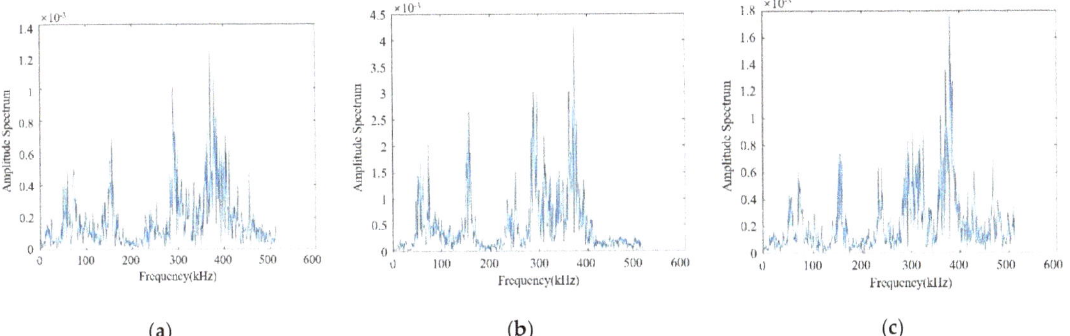

(a) (b) (c)

Figure 10. Frequency domain waveform of the P-signals of each sample. (**a**) A1.No.17715, (**b**) B1.No.8200, (**c**) C1.No.6375.

This phenomenon shows that the selection of the P-signals is not strongly affected by the test environment but is mainly determined by the physical properties of rock. This means that the P-signals defined in this paper have strong representativeness and applicability.

5.2. Wavelet Packet Energy Analysis of AE Signals

Traditional analysis methods such as the FFT are not suitable to analyze typical non-stationary signals such as AE signals owing to the defects of the algorithm itself. By contrast, wavelet analysis is a localized analysis in the time and frequency domains, and it can not only present the frequency-domain analysis of the signal in local time but also describe the time-domain information corresponding to the frequency-domain information. Based on wavelet analysis theory, wavelet packet analysis introduces the concept of optimal basis selection, which has higher time–frequency resolution than wavelet analysis [37]. Therefore, wavelet packet analysis has greater advantages in the time–frequency domain analysis, feature extraction, and AE signal energy analysis.

The concept of the frequency centroid described in the previous section clarifies that the frequency centroid will be close to the dominant energy band of the signal. Therefore, in this section, we use wavelet packet analysis to further analyze the energy of the P-signals to obtain the energy distribution characteristics of the P-signals in each band. More importantly, we verify the practicability of peak stress prediction by the P-signals.

5.2.1. Wavelet Packet Frequency Band Decomposition of the P-Signals

The AE signals generated in the loading process of a rock sample are abruptly occurring non-stationary signals with complex characteristics. Wavelet packet analysis can effectively identify these signals and further decompose their high-frequency and low-frequency components. Then, the corresponding frequency band is adaptively selected according to the characteristics of the decomposed signals. According to the sampling law, if the sampling frequency of the AE system is 1 MHz, the Nyquist frequency is 500 kHz. Rock is a type of anisotropic and heterogeneous material, with random microstructural planes. Different wavelet bases provide different results. Various wavelet bases have been applied to wavelet packet analysis, such as Symlets basis, Morlet basis, and Meyer basis [38–42]. In view of the complexity and diversity of AE signals, wavelet bases with symmetry and tight support should be selected [43,44]. In this paper, the db3 wavelet basis function of Daubechies wavelet series was selected and the AE signals were decomposed to the fourth layer. A total of 16 sub-bands (labeled 1–16) with a band length of 31.25 kHz were obtained. The specific range of each frequency band is shown in Table 2.

Table 2. Corresponding labels to band ranges.

Sub-Band Label	Frequency Band Range/kHz
1	0~31.25
2	31.25~62.5
3	62.5~93.75
. . .	. . .
16	478.75~500

5.2.2. Energy Distribution of Each Sub-Band

When the AE signal is decomposed to the fourth layer, the analyzed signal is recorded as $S_{4,j}$, and its corresponding energy is $E_{4,j}$; then, the energy of each frequency band can be expressed as

$$E_{4,j} = \int \left| S_{4,j}(t) \right|^2 dt = \sum_{k=1}^{m} \left| x_{j,k} \right|^2 \tag{5}$$

where m is the number of discrete sampling points of the signal; $x_{j,k}$ represents the amplitude of $S_{4,j}$ discrete points of the signal.

That is, if the total energy of the analyzed signal is E_0, then

$$E_0 = \sum_{j=0}^{2^4-1} E_{4,j} \tag{6}$$

The percentage of the energy of each sub-band of the signal in the total energy of the analyzed signal is calculated as follows:

$$P_j = \frac{E_{4,j}}{E_0} \tag{7}$$

Based on the above formula, the energy ratio distribution of the AE signal in each sub-band after wavelet packet decomposition can be obtained.

5.2.3. Energy Distribution of the P-Signals

In Section 5.1, the frequency-domain characteristics of the P-signals obtained by FFT are extracted. In this section, the P-signals of granite samples under each confining pressure are extracted by the wavelet packet energy analysis method, and the proportions of all P-signals in the sub-bands of granite samples under various confining pressures are obtained, as shown in Figure 11.

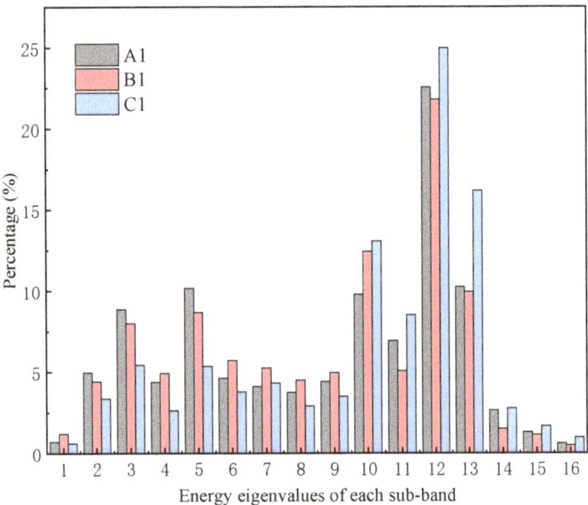

Figure 11. Percentage of energy in each frequency band of the P-signals.

Figure 11 shows that the frequency band energy of the P-signals under various confining pressures is mainly concentrated in the No. 10–13 frequency band (281.25–406.25 kHz), accounting for more than 50% of the total signals. This result confirms the rationality of defining the signal with frequency centroid greater than 275 kHz as the P-signal. The No. 12 frequency band (343.75–375 kHz) accounts for the highest proportion of the single frequency band. This result validates the definition of the P-signal as the signal with a peak frequency greater than 375 kHz. The No. 10 frequency band (281.25–312.5 kHz) occupies the second-highest proportion of all frequency bands because the second peak frequency of most signals is concentrated in this area. However, the energy proportion of the single frequency band is less than 25%. As the confining pressure increases, the energy proportion of low-frequency signal decreases.

The energy proportions of all the P-signals under various confining pressure values in each frequency band obtained through the wavelet packet analysis are the same as the conditions proposed in the previous section for defining P-signals. That is, the results confirm that the peak frequency and frequency centroid of the precursor signals should be greater than 350 and 275 kHz, respectively.

6. Discussion

The geological conditions of deep well construction engineering are complex, and the surrounding rocks are prone to rockburst, spalling, and other geological disasters; these conditions are quite different from those of shallow rocks. The deformation characteristics and mechanical behavior of deep rock under high confining pressure serve as effective indicators for the construction safety of deep shaft structures. As a non-destructive testing technology, the use of the AE signal monitoring technology to determine the deformation and failure characteristics of rock has been extensively studied. This method works on the principle that the damage and failure of brittle rock are attributed to the process of the initial compaction and closure of internal microcracks, followed by the nucleation of the microcracks and gradual development of large-scale cracks, resulting in macroscopic failure [30]. Therefore, this paper focused on the physical and mechanical properties of granite in the deep strata of Sanshandao Gold Mine in Laizhou. The loading failure process of deep granite was first divided into five distinct stages according to the AE IET function. Then, test results demonstrated that the deep granite shows extremely strong brittle characteristics under different confining pressures in the triaxial test.

In practical engineering, the surrounding rock stress field is redistributed after the excavation of deep rock mass, resulting in regional stress concentration and other phenomena. If the appropriate support is not applied in a timely manner, the rock mass will eventually be damaged. In this paper, the effect of rock depth on rock strength, deformation, and AE characteristics was further revealed by the triaxial tests under different confining pressures. The test results showed that the deep granite has higher compressive strength and elastic brittleness than the shallow rock [3,4]. For the granite samples with lower confining pressure, under the same loading conditions, first, the macroscopic fracture is generated and energy is increased, and the phenomenon of high AE rate occurs in the early and middle stages. This indicates that the development of internal cracks in rock is more active under low confining pressure. Under high confining pressure, some natural pores and microcracks are compacted in advance by the confining pressure, so the AE rate is extremely low in the early stage. High confining pressure also results in the accumulation of a large amount of energy in the stage of macroscopic fracture generation, and energy is rapidly released in the post-peak stage at an extremely high rate. These results suggest that the greater the buried depth, the higher the strength and elastic brittleness of the rock, and the redistribution process of the stress field after the excavation of the deep well causes different mechanical responses to the rock in the area.

In this study, we analyzed the frequency-domain characteristics of AE signals in the entire loading process to determine the early warning AE signals, which can reveal the critical state of rock failure. During the loading process, the rock undergoes microcrack nucleation and crack growth, followed by the formation of large-scale cracks and a fracture surface, eventually resulting in rock failure [30]. To better describe the AE signal when the rock reaches the peak stress, the relationship of the peak frequency, amplitude, and centroid frequency with axial stress was studied under the test conditions. The P-signals related to the frequency-domain characteristics were defined exactly. The test results showed that the peak stress of rock rapidly reaches the critical point of loading after a large number of P-signals appear. Next, the energy distribution of the P-signals was studied accurately by wavelet packet analysis, and the feasibility of the P-signal definition was verified. According to the inverse relationship between AE frequency and crack size [29,41], it can be inferred that with increasing confining pressure, the proportion of small cracks in granite increases.

However, in practical engineering, environments with high temperature and high pore water pressure clearly affect rock's physical and mechanical properties as well as AE characteristics. In this study, the high confining pressure environment was simulated under in situ conditions. To further study the effect of high temperature and high pore water pressure on the mechanical properties of rock and the characteristics of precursory AE of rock under the coupling conditions of multiple factors closer to the actual engineering environment, it is necessary to conduct in-depth processing and analysis using a true triaxial testing machine with THMC coupling effects in the future.

7. Conclusions

This study systematically investigated the damage and failure evolution law and frequency-domain distribution characteristics of AE signals of deep granite under tri-axial loading tests. Further, the conditions for the prediction signals before peak stress were defined. The important conclusions are summarized as follows:

(1) The AE IET function could effectively identify the deformation and failure degree of rocks, and the loading process was divided into five stages. In the early stage of loading, the function fluctuated in a low region; however, the rapid increase in its value in the later stage indicates the generation of macroscopic cracks and rock fracture.

(2) The AE signals with amplitude greater than 85 dB, peak frequency greater than 350 kHz, and frequency centroid greater than 275 kHz were defined as the precursor signals before the occurrence of the peak stress. The signals are applicable to all samples under different confining pressures and have good universality and applicability.

(3) Through wavelet packet analysis, the frequency domain was divided into 16 sub-frequency bands of 31.25 kHz. The energy proportion of the characteristic signal was mainly concentrated in the 343.75–375 kHz frequency band, followed by the 281.25–312.5 kHz frequency band, and these were related to the distribution of the primary peak frequency and the secondary peak frequency of all signals. These results agree well with the selection standard of the precursor signals.

Author Contributions: Conceptualization, H.J.; methodology, L.Z.; software, L.Z. and L.L.; validation, J.Z.; writing—original draft preparation, L.Z.; writing—review and editing, L.L.; supervision, H.J. All authors have read and agreed to the published version of the manuscript.

Funding: This research was funded by the Beijing Natural Science Foundation (No. 2204084), the National Key Research Development Program of China (No. 2016YFC0600801), and the National Natural Science Foundation of China (Nos. 51534002 and 52004015).

Conflicts of Interest: The authors declare no conflict of interest.

References

1. Feng, X.-T.; Liu, J.; Chen, B.; Xiao, Y.; Feng, G.-L.; Zhang, F. Monitoring, Warning, and Control of Rockburst in Deep Metal Mines. *Engineering* **2017**, *3*, 538–545. [CrossRef]
2. Cai, M.; Brown, E.T. Challenges in the Mining and Utilization of Deep Mineral Resources. *Engineering* **2017**, *3*, 432–433. [CrossRef]
3. Li, X.; Gong, F.; Tao, M.; Dong, L.; Du, K.; Ma, C.; Zhou, Z.; Yin, T. Failure mechanism and coupled static-dynamic loading theory in deep hard rock mining: A review. *J. Rock Mech. Geotech. Eng.* **2017**, *9*, 767–782. [CrossRef]
4. Gong, Q.M.; Yin, L.J.; Wu, S.Y.; Zhao, J.; Ting, Y. Rock burst and slabbing failure and its influence on TBM excavation at headrace tunnels in Jinping II hydropower station. *Eng. Geol.* **2012**, *124*, 98–108. [CrossRef]
5. Martin, C.D.; Christiansson, R. Estimating the potential for spalling around a deep nuclear waste repository in crystalline rock. *Int. J. Rock Mech. Min. Sci.* **2009**, *46*, 219–228. [CrossRef]
6. Armaghani, D.J.; Mamou, A.; Maraveas, C.; Roussis, P.C.; Siorikis, V.G.; Skentou, A.D.; Asteris, P.G. Predicting the unconfined compressive strength of granite using only two non-destructive test indexes. *Geomech. Eng.* **2021**, *25*, 317–330.
7. Du, K.; Liu, M.; Yang, C.; Tao, M.; Wang, S. Mechanical and Acoustic Emission (AE) Characteristics of Rocks under Biaxial Confinements. *Appl. Sci.* **2021**, *11*, 769. [CrossRef]
8. Li, H.; Dong, Z.; Yang, Y.; Liu, B.; Chen, M.; Jing, W. Experimental Study of Damage Development in Salt Rock under Uniaxial Stress Using Ultrasonic Velocity and Acoustic Emissions. *Appl. Sci.* **2018**, *8*, 553. [CrossRef]
9. Liu, X.; Zhang, H.; Wang, X.; Zhang, C.; Xie, H.; Yang, S.; Lu, W. Acoustic Emission Characteristics of Graded Loading Intact and Holey Rock Samples during the Damage and Failure Process. *Appl. Sci.* **2019**, *9*, 1595. [CrossRef]
10. Cai, M.; Morioka, H.; Kaiser, P.; Tasaka, Y.; Kurose, H.; Minami, M.; Maejima, T. Back-analysis of rock mass strength parameters using AE monitoring data. *Int. J. Rock Mech. Min. Sci.* **2006**, *44*, 538–549. [CrossRef]
11. Du, K.; Li, X.; Tao, M.; Wang, S. Experimental study on acoustic emission (AE) characteristics and crack classification during rock fracture in several basic lab tests. *Int. J. Rock Mech. Min. Sci.* **2020**, *133*, 104411. [CrossRef]
12. Peng, S.; Sbartaï, Z.M.; Parent, T. Mechanical damage evaluation of masonry under tensile loading by acoustic emission technique. *Constr. Build. Mater.* **2020**, *258*, 120336. [CrossRef]
13. Lei, X.; Satoh, T. Indicators of critical point behavior prior to rock failure inferred from pre-failure damage. *Tectonophysics* **2007**, *431*, 97–111. [CrossRef]
14. Zhang, A.; Zhang, R.; Gao, M.; Zhang, Z.; Jia, Z.; Zhang, Z.; Zha, E. Failure Behavior and Damage Characteristics of Coal at Different Depths under Triaxial Unloading Based on Acoustic Emission. *Energies* **2020**, *13*, 4451. [CrossRef]
15. Zhang, J. Investigation of Relation between Fracture Scale and Acoustic Emission Time-Frequency Parameters in Rocks. *Shock. Vib.* **2018**, *2018*, 1–14. [CrossRef]
16. Liu, J.; Wu, N.; Si, G.; Zhao, M. Experimental study on mechanical properties and failure behaviour of the pre-cracked coal-rock combination. *Bull. Int. Assoc. Eng. Geol.* **2020**, *80*, 2307–2321. [CrossRef]
17. Tuncay, E.; Obara, Y. Comparison of stresses obtained from Acoustic Emission and Compact Conical-Ended Borehole Overcoring techniques and an evaluation of the Kaiser Effect level. *Bull. Int. Assoc. Eng. Geol.* **2011**, *71*, 367–377. [CrossRef]
18. Carpinteri, A.; Lacidogna, G.; Niccolini, G.; Puzzi, S. Critical defect size distributions in concrete structures detected by the acoustic emission technique. *Meccanica* **2007**, *43*, 349–363. [CrossRef]
19. Carpinteri, A.; Lacidogna, G.; Puzzi, S. From criticality to final collapse: Evolution of the "b-value" from 1.5 to 1.0. *Chaos Soliton Fract.* **2009**, *41*, 843–853. [CrossRef]
20. Sagar, R.V.; Rao, M. An experimental study on loading rate effect on acoustic emission based b-values related to reinforced concrete fracture. *Constr. Build. Mater.* **2014**, *70*, 460–472. [CrossRef]
21. Dong, L.; Zhang, Y.; Ma, J. Micro-Crack Mechanism in the Fracture Evolution of Saturated Granite and Enlightenment to the Precursors of Instability. *Sensors* **2020**, *20*, 4595. [CrossRef]

22. Ohtsu, M.; Tomoda, Y. Corrosion Process in Reinforced Concrete Identified by Acoustic Emission. *Mater. Trans.* **2007**, *48*, 1184–1189. [CrossRef]
23. Rodríguez, P.; Celestino, T.B. Application of acoustic emission monitoring and signal analysis to the qualitative and quantitative characterization of the fracturing process in rocks. *Eng. Fract. Mech.* **2019**, *210*, 54–69. [CrossRef]
24. Xu, X.; Dou, L.; Lu, C.; Zhang, Y. Frequency spectrum analysis on micro-seismic signal of rock bursts induced by dynamic disturbance. *Int. J. Min. Sci. Technonol.* **2010**, *20*, 682–685. [CrossRef]
25. Aggelis, D.; Mpalaskas, A.; Matikas, T. Acoustic signature of different fracture modes in marble and cementitious materials under flexural load. *Mech. Res. Commun.* **2013**, *47*, 39–43. [CrossRef]
26. He, M.; Miao, J.; Feng, J. Rock burst process of limestone and its acoustic emission characteristics under true-triaxial unloading conditions. *Int. J. Rock Mech. Min. Sci.* **2010**, *47*, 286–298. [CrossRef]
27. Li, L.R.; Deng, J.H.; Zheng, L.; Liu, J.F. Dominant Frequency Characteristics of Acoustic Emissions in White Marble during Direct Tensile Tests. *Rock Mech. Rock Eng.* **2017**, *50*, 1337–1346. [CrossRef]
28. Zhang, Y.; Wu, W.; Yao, X.; Liang, P.; Sun, L.; Liu, X. Study on Spectrum Characteristics and Clustering of Acoustic Emission Signals from Rock Fracture. *Circuits Syst. Signal Process.* **2019**, *39*, 1133–1145. [CrossRef]
29. Wang, Y.; Han, J.; Li, C. Acoustic emission and CT investigation on fracture evolution of granite containing two flaws subjected to freeze–thaw and cyclic uniaxial increasing-amplitude loading conditions. *Constr. Build. Mater.* **2020**, *260*, 119769. [CrossRef]
30. Ohnaka, M.; Mogi, K. Frequency characteristics of acoustic emission in rocks under uniaxial compression and its relation to the fracturing process to failure. *J. Geophys Res.* **1982**, *87*, 3873–3884. [CrossRef]
31. Moradian, Z.; Einstein, H.H.; Ballivy, G. Detection of Cracking Levels in Brittle Rocks by Parametric Analysis of the Acoustic Emission Signals. *Rock Mech. Rock Eng.* **2015**, *49*, 785–800. [CrossRef]
32. Xue, L.; Qin, S.; Sun, Q.; Wang, Y.; Qian, H. A quantitative criterion to describe the deformation process of rock sample subjected to uniaxial compression: From criticality to final failure. *Phys. A Stat. Mech. Its Appl.* **2014**, *410*, 470–482. [CrossRef]
33. Zhang, J.Z.; Zhou, X.P.; Zhou, L.S.; Berto, F. Progressive failure of brittle rocks with non-isometric flaws: Insights from acousto-optic-mechanical (AOM) data. *Fatigue Fract. Eng. Mater.Structures* **2019**, *42*, 1787–1802. [CrossRef]
34. Zhou, X.-P.; Zhang, J.-Z.; Qian, Q.-H.; Niu, Y. Experimental investigation of progressive cracking processes in granite under uniaxial loading using digital imaging and AE techniques. *J. Struct. Geol.* **2019**, *126*, 129–145. [CrossRef]
35. Lennartz-Sassinek, S.; Main, I.G.; Zaiser, M.; Graham, C.C. Acceleration and localization of subcritical crack growth in a natural composite material. *Phys. Rev. E* **2014**, *90*, 052401. [CrossRef] [PubMed]
36. Triantis, D.; Kourkoulis, S.K. An Alternative Approach for Representing the Data Provided by the Acoustic Emission Technique. *Rock Mech. Rock Eng.* **2018**, *51*, 2433–2438. [CrossRef]
37. Tang, S.; Tong, M.; Hu, J.; He, X. Characteristics of acoustic emission signals in damp cracking coal rocks. *Int. J. Min. Sci Technol.* **2010**, *20*, 143–147. [CrossRef]
38. Ebrahimian, Z.; Ahmadi, M.; Sadri, S.; Li, B.; Moradian, O. Wavelet analysis of acoustic emissions associated with cracking in rocks. *Eng. Fract. Mech.* **2019**, *217*, 106516. [CrossRef]
39. Lee, I.-M.; Han, S.-I.; Kim, H.-J.; Yu, J.-D.; Min, B.-K.; Lee, J.-S. Evaluation of rock bolt integrity using Fourier and wavelet transforms. *Tunn. Undergr. Space Technol.* **2012**, *28*, 304–314. [CrossRef]
40. Liu, X.; Liang, Z.; Zhang, Y.; Wu, X.; Liao, Z. Acoustic Emission Signal Recognition of Different Rocks Using Wavelet Transform and Artificial Neural Network. *Shock. Vib.* **2015**, *2015*, 1–14. [CrossRef]
41. Liu, X.; Wu, L.; Zhang, Y.; Liang, Z.; Yao, X.; Liang, P. Frequency properties of acoustic emissions from the dry and saturated rock. *Environ. Earth Sci.* **2019**, *78*, 67. [CrossRef]
42. Liu, X.-L.; Liu, Z.; Li, X.-B.; Rao, M.; Dong, L.-J. Wavelet threshold de-noising of rock acoustic emission signals subjected to dynamic loads. *J. Geophys. Eng.* **2018**, *15*, 1160–1170. [CrossRef]
43. Wang, Y.; Chen, S.J.; Liu, S.J.; Hu, H.X. Best wavelet basis for wavelet transforms in acoustic emission signals of concrete damage process. *Russ. J. Nondestruct. Test.* **2016**, *52*, 125–133. [CrossRef]
44. Wang, Z.; Ning, J.; Ren, H. Frequency characteristics of the released stress wave by propagating cracks in brittle materials. *Theor. Appl. Fract. Mech.* **2018**, *96*, 72–82. [CrossRef]

MDPI

St. Alban-Anlage 66

4052 Basel

Switzerland

Tel. +41 61 683 77 34

Fax +41 61 302 89 18

www.mdpi.com

Applied Sciences Editorial Office

E-mail: applsci@mdpi.com

www.mdpi.com/journal/applsci